D1985597

MATHEMATICAL SURVEYS

NUMBER I

# THE PROBLEM OF MOMENTS

BY

J. A. SHOHAT
UNIVERSITY OF PENNSYLVANIA

AND

J. D. TAMARKIN
BROWN UNIVERSITY

PUBLISHED BY THE
AMERICAN MATHEMATICAL SOCIETY
Providence, Rhode Island 02940
1943

Fourth printing of revised edition, 1970
(Revised edition issued in 1950)

The revised edition included revisions, corrections and a Supplementary Bibliography. The third printing included a minor correction and an addition to the Bibliography. This fourth edition is merely a reissue of the third edition.

International Standard Book Number 0-8218-1501-6
Library of Congress Catalog Card Number 51-96





PRINTED IN THE UNITED STATES OF AMERICA

# TABLE OF CONTENTS

# PREFACE

The problem of moments is of fairly old origin, but it received its first systematic treatment in the works of Tchebycheff, Markoff, Stieltjes, and, later, Hamburger, Nevanlinna, M. Riesz, Hausdorff, Carleman, and Stone. The subject has an extensive literature, but has not been treated in book or monograph form. In view of the considerable mathematical (and also practical) interest of the moment problem it appeared to the authors desirable to submit such a treatment to a wide mathematical public. In the present monograph the main attention is given to the classical moment problem, and, with the exception of a few remarks concerning the trigonometrical moment problem, no mention is made of various generalizations and modifications, important as they may be. Furthermore, lack of space did not permit the treatment of important developments of Carleman and Stone based on the theory of singular integral equations and operators in Hilbert space. On the other hand, a special chapter is devoted to the theory of approximate (mechanical) quadratures, which is intimately related to the theory of moments and in many instances throws additional light on the situation.

The bibliography at the end of the book makes no claim to completeness.

The authors wish to acknowledge with gratitude the help received from Brown University and the University of Pennsylvania (through its Faculty Research Committee) in preparing this manuscript for publication.

In the second printing the theorem of §4, page xiii, has been corrected by R. P. Boas, and a supplementary bibliography has been added.

# INTRODUCTION

**1. Brief historical review.** In 1894–95 Stieltjes published a classical paper: "Recherches sur les fractions continues" containing a wealth of new ideas; among others, a new concept of integral—our modern "Stieltjes Integral". In this paper he proposes and solves completely the following problem which he calls "Problem of Moments":

Find a bounded non-decreasing function $\psi(x)$ in the interval $[0, \infty)$ such that its "moments" $\int_0^\infty x^n\, d\psi(x)$, $n = 0, 1, 2, \cdots$, have a prescribed set of values

$$(1) \qquad \int_0^\infty x^n\, d\psi(x) = \mu_n, \qquad n = 0, 1, 2, \cdots.$$

The terminology "Problem of Moments" is taken by Stieltjes from Mechanics. [Stieltjes uses on many occasions mechanical concepts of mass, stability, etc., in solving analytical problems.] If we consider $d\psi(x)$ as a mass distributed over $[x, x + dx]$ so that $\int_0^x d\psi(t)$ represents the mass distributed over the segment $[0, x]$—whence the modern designation of $\psi(x)$ as "distribution function"—then $\int_0^\infty x\, d\psi(x)$, $\int_0^\infty x^2\, d\psi(x)$ represent, respectively, the first (statical) moment and the second moment (moment of inertia) with respect to 0 of the total mass $\int_0^\infty d\psi(x)$ distributed over the real semi-axis $[0, \infty)$. Generalizing, Stieltjes calls $\int_0^\infty x^n\, d\psi(x)$ the $n$-th moment, with respect to 0, of the given mass distribution characterized by the function $\psi(x)$.

Stieltjes makes the solution of the Moment-Problem (1) dependent upon the nature of the continued fraction "corresponding" to the integral

$$(2) \qquad \begin{aligned} I(z, \psi) = \int_0^\infty \frac{d\psi(y)}{z + y} &\sim \frac{\mu_0}{z} - \frac{\mu_1}{z^2} + \frac{\mu_2}{z^3} - \frac{\mu_3}{z^4} + \cdots \\ &\sim \frac{1|}{|a_1 z} + \frac{1|}{|a_2} + \frac{1|}{|a_3 z} + \frac{1|}{|a_4} + \cdots, \end{aligned}$$

and upon the closely related "associated" continued fraction

$$(3) \qquad \frac{\lambda_1|}{|z + c_1} - \frac{\lambda_2|}{|z + c_2} - \frac{\lambda_3|}{|z + c_3} - \cdots$$

derived from (2) by "contraction":

$$z - \frac{\alpha|}{|1} - \frac{\beta|}{|z - \gamma} = z - \alpha - \frac{\alpha\beta}{z - (\beta + \gamma)}.$$

Making use of the theory of continued fractions Stieltjes shows that in (2) all $a_i$ are positive (which results in the positiveness of all $\lambda_i$ and $c_i$ in (3)).*

He further shows that this necessary condition is also sufficient for the existence of a solution of the Problem of Moments (1). In terms of the given sequence $\{\mu_n\}$ this condition is equivalent to the positiveness of the following determinants

$$(4)\qquad \Delta_n = \begin{vmatrix} \mu_0 & \mu_1 & \cdots & \mu_n \\ \mu_1 & \mu_2 & \cdots & \mu_{n+1} \\ \cdots & \cdots & \cdots & \cdots \\ \cdots & \cdots & \cdots & \cdots \\ \mu_n & \mu_{n+1} & \cdots & \mu_{2n} \end{vmatrix} \equiv |\, \mu_{i+j} \,|_{i,j=0}^{n}; \qquad n = 0, 1, 2, \cdots,$$

$$(5)\qquad \Delta_n^{(1)} = \begin{vmatrix} \mu_1 & \mu_2 & \cdots & \mu_n & \mu_{n+1} \\ \mu_2 & \mu_3 & \cdots & \mu_{n+1} & \mu_{n+2} \\ \cdots & \cdots & \cdots & \cdots & \cdots \\ \cdots & \cdots & \cdots & \cdots & \cdots \\ \mu_{n+1} & \mu_{n+2} & \cdots & \mu_{2n} & \mu_{2n+1} \end{vmatrix} \equiv |\, \mu_{i+j+1} \,|_{i,j=0}^{n}; \qquad n = 0, 1, 2, \cdots.$$

The solution may be unique, in which case we speak of a "determined Moment-Problem"; or there may be more than one solution in which case there are, of necessity, infinitely many solutions; our Moment-Problem is then "indeterminate". Stieltjes illustrates the latter case by a remarkable example. He further gives an effective construction of certain solutions of the Moment-Problem (all, of course, essentially the same in case of a determined problem) which in the indeterminate case turn out to possess important minimal properties. Here the denominators of the successive approximants to the continued fractions (2) and (3) play an important role. In passing Stieltjes introduces an important new proposition dealing with the convergence of series of functions of a complex variable (now known as the Stieltjes-Vitali Theorem) which leads to a complete solution of the problem of convergence of the continued fraction (2) in the complex $z$-plane. Here Stieltjes shows that the Moment-Problem (1) is determined or indeterminate according as the continued fraction (2) is convergent or divergent, that is, according as the series $\sum_1^\infty a_i$ diverges or converges. The interesting fact that the continued fraction (2) may converge for certain $z$ (to the value $I(z, \psi)$), while the series $\sum_0^\infty (-1)^i \mu_i z^{-i-1}$ diverges for all $z$ is demonstrated.

* In the subsequent discussion we write

$$\int_0^\infty \frac{d\psi(y)}{z-y} \sim \frac{\mu_0}{z} + \frac{\mu_1}{z^2} + \frac{\mu_2}{z^3} + \cdots,$$

so that the corresponding and associated continued fractions (2) and (3) are replaced respectively by

$$\frac{1\,|}{|\,l_1 z} + \frac{1\,|}{|\,l_2} + \frac{1\,|}{|\,l_3 z} + \cdots \quad \text{and} \quad \frac{\lambda_1\,|}{|\,z - c_1} - \frac{\lambda_2\,|}{|\,z - c_2} - \frac{\lambda_3\,|}{|\,z - c_3} - \cdots,$$

where all $l_{2i+1}$ are positive, all $l_{2i}$ are negative, and $\lambda_i$ and $c_i$ are positive.

Stieltjes was not the first to discuss either the Moment-Problem, or the continued fraction (3). The first considerations along these lines are due to the great Russian mathematician Tchebycheff who in a series of papers started in 1855 discusses integrals of the type $\int_{-\infty}^{\infty} \frac{p(y)\,dy}{x-y}$, where $p(x)$ is non-negative in $(-\infty, \infty)$, and sums of the type $\sum_{-\infty}^{\infty} \frac{\theta_i^2}{x - x_i}$, $\theta_i \neq 0$ (both cases are now covered by a Stieltjes Integral). Tchebycheff's main tool is the theory of continued fractions which he uses with extreme ingenuity. However, Tchebycheff was not interested in the existence or construction of a solution of the Moment-Problem,

$$\int_{-\infty}^{\infty} p(x)x^n\,dx = \mu_n, \qquad n = 0, 1, 2, \cdots, \tag{6}$$

but mainly in the following two problems: a) How far does a given sequence of moments determine the function $p(x)$? More particularly, given

$$\int_{-\infty}^{\infty} p(x)x^n\,dx = \int_{-\infty}^{\infty} e^{-x^2}x^n\,dx, \qquad n = 0, 1, 2, \cdots;$$

can we conclude that $p(x) = e^{-x^2}$, or, as we say now, that the distribution characterized by the function $\int_{-\infty}^{x} p(t)\,dt$ is a normal one? This is a fundamental problem in the theory of probability and in mathematical statistics. b) What are the properties of the polynomials $\omega_n(z)$, denominators of successive approximants to the continued fraction (3)? This opened a vast new field, the general theory of orthogonal polynomials, of which only the classical polynomials of Legendre, Jacobi, Abel—Laguerre and Laplace—Hermite were known before Tchebycheff. In the work of Tchebycheff we find numerous applications of orthogonal polynomials to interpolation, approximate quadratures, expansion of functions in series. Later they have been applied to the general theory of polynomials, theory of best approximations, theory of probability and mathematical statistics.

Another work which preceded that of Stieltjes is the classical work of Heine (1861, 1878, 1881). Here we find a brief discussion of the continued fraction associated with $\int_a^b \frac{p(y)\,dy}{x-y}$, where the given function $p(x)$ is non-negative in $(a, b)$, and also an application of the orthogonal polynomials $\omega_n(x)$ to approximate quadratures.

One may venture the opinion that the use of this integral and of continued fractions was suggested to Stieltjes by the work of Tchebycheff and Heine. We must emphasize the importance of the new analytical tool, the Stieltjes Integral, which made it possible to treat the Problem of Moments in its most general form, namely,

$$\int_{-\infty}^{\infty} x^n\,d\psi(x) = \mu_n, \qquad n = 0, 1, 2, \cdots. \tag{7}$$

One of the most talented pupils of Tchebycheff, A. Markoff, continued the work of his teacher applying it, in particular, to the theory of probability ("method of moments" applied to the proof of the fundamental limit-theorem), and to the closely related problem of finding precise bounds for $\int_c^d d\psi(x)$, $a < c < d < b$, where the function $\psi(x)$ is non-decreasing in $(a, b)$, its first $n + 1$ moments being given. This important problem was proposed and its solution, based on the now celebrated "Tchebycheff inequalities", was given without proof by Tchebycheff in 1873. The proof was supplied by Markoff in his Thesis in 1884. It is interesting to note that Tchebycheff inequalities have been proved simultaneously and in the same manner by Stieltjes. Markoff further generalizes the Moment-Problem (1896) by requiring the solution $p(x)$ to be bounded:

$$\int_{-\infty}^{\infty} x^n p(x)\, dx = \mu_n, \qquad n = 0, 1, 2, \cdots, \quad \text{with} \quad 0 \leqq p(x) \leqq L.$$

In his investigations, as in those of his teacher, continued fractions play a predominant role.

As often happens in the history of science, the Problem of Moments lay dormant for more than 20 years. It revived again in the work of H. Hamburger, R. Nevanlinna, M. Riesz, T. Carleman, Hausdorff, and others.

An important approach to, and extension of, the work of Stieltjes to the whole real axis $(-\infty, \infty)$ was achieved by H. Hamburger (1920, 1921). This extension is by no means trivial. The consideration of negative values of $x$ introduces new factors in the situation. Hamburger makes extensive use of Helly's theorem of choice. He fully discusses the convergence in the complex plane of both the associated and the corresponding (if it exists) continued fractions. He shows that a necessary and sufficient condition for the existence of a solution of the Moment-Problem (7) is the positiveness of all determinants $\Delta_n$ in (4), and also gives criteria for the Moment-Problem (7) to be determined or indeterminate. A curious fact is revealed, namely, that the Moment-Problem (7) may be indeterminate while the corresponding Stieltjes Moment-Problem (1), with the same $\mu_n$, is determined.

R. Nevanlinna (1922) makes use of the modern theory of functions and exhibits the solutions of the Moment-Problem (7) and their properties in terms of the functions

$$I(z; \psi) = \int_{-\infty}^{\infty} \frac{d\psi(y)}{z - y}, \qquad z \text{ complex.}$$

To him is due the important notion of "extremal solutions".

About the same time (1921, 1922, 1923) M. Riesz solved the Moment-Problem (7) on the basis of "quasi-orthogonal polynomials", i.e. linear combinations $A_n\omega_n(z) + A_{n-1}\omega_{n-1}(z)$. He also showed the close connection between the Problem of Moments and the so-called "closure property" (Parseval Formula) for the orthogonal polynomials $\omega_n(z)$.

Carleman (1923, 1926) shows the connection between the Problem of Moments (7) and the theories of quasi-analytic functions and of quadratic forms in infinitely many variables (through the medium of the asymptotic series $\sum_0^\infty \mu_i z^{-i-1}$). To him is due the most general criterion, so far known, for the Moment-Problem to be determined.

Hausdorff (1923) gives criteria for the Moment-Problem (7) to possess a (necessarily unique) solution in a finite interval, that is, when $\psi(x)$ in (7) is required to remain constant outside a given finite interval. An effective construction of the solution is given and criteria are derived for the solution to have specified properties—continuity, differentiability, etc.

The interest in the Problem of Moments remains strong up to the present day. Among the most important contributions we may mention the work of Achyeser and Krein (1934). They have generalized the work of Markoff, making use of the tools of the theory of quadratic forms; they also extended the theory to the "trigonometric Moment-Problem"

$$(8) \qquad \int_0^{2\pi} e^{inx}\, d\psi(x) = \mu_n, \qquad n = 0, 1, 2, \cdots.$$

Compare, in this connection, the work of S. Verblunsky (1932).

Carleman, and later, Stone developed a rather complete treatment of the Moment-Problem on the basis of the theory of Jacobi quadratic forms and singular integral equations and operators in Hilbert space. Finally, Haviland and Cramér extended M. Riesz' theory to the case of several dimensions.

Various generalizations have been made to the cases where the set of functions $\{x^n\}$ is replaced by a more general set $\{\varphi_n(x)\}$, or the integrals by more general linear operators in abstract spaces. These generalizations, however, will not be considered in the present monograph.

The discussion in the first two chapters follows the work of M. Riesz and R. Nevanlinna, in chapter III that of Markoff, Achyeser and Krein, and Hausdorff.

We shall now state explicitly, but mostly without proof, some fundamental facts which will be used in various places in the following chapters.

**2. Distribution functions.** Let $\mathfrak{R}_k$ be a $k$-dimensional Euclidean space. A function $\Phi(e)$ of sets $e$ in $\mathfrak{R}_k$ is called a *distribution set-function* if it is non-negative, defined (and finite) over the family of all Borel sets in $\mathfrak{R}_k$, and is completely additive:

$$\sum_{i=1}^{\infty} \Phi(e_i) = \Phi\left(\sum_{i=1}^{\infty} e_i\right), \qquad e_i e_j = 0, \quad \text{if} \quad i \neq j.$$

The *spectrum* $\mathfrak{S}(\Phi)$ of a distribution set-function $\Phi(e)$ is defined as the set of all points $x \in \mathfrak{R}_k$, such that $\Phi(G) > 0$ for every open set $G$ containing $x$.

The *point spectrum* of $\Phi$ is the set of all points $x$ such that $\Phi((x)) > 0$.

By an interval $I \subset \mathfrak{R}_k$ we mean the set of points $x = (x_1, x_2, \cdots, x_k)$ whose coordinates satisfy conditions $a_i < x_i \leqq b_i$, $i = 1, 2, \cdots, k$, with obvious modifications in the case of an open or closed interval.

An interval $I$ is an *interval of continuity* of the distribution set-function $\Phi$ (or more generally, for an additive function $\Phi(I)$ defined over all intervals) if, on introducing

$$I_\delta^\pm: \quad a_i \mp \delta < x_i \leqq b_i \pm \delta, \qquad i = 1, 2, \cdots, k,$$

we have

$$\Phi(I_\delta^\pm) \to \Phi(I), \qquad \text{as} \quad \delta \to 0.$$

Two distribution set-functions are said to be *substantially equal* if they have the same intervals of continuity and their values coincide over all such intervals.

Let $\Psi(I)$ be a non-negative set-function defined (and finite) over all intervals $I$ in $\Re_k$ and satisfying the condition

$$\Psi(I) \leqq \sum_{i=1}^{n} \Psi(I_i), \quad \text{whenever} \quad I = \sum_{i=1}^{n} I_i, \qquad I_i I_j = 0 \quad \text{for} \quad i \neq j.$$

It is always possible to extend $\Psi(I)$ to a distribution set-function $\Phi(e)$ defined at least for all Borel sets having the same intervals of continuity as $\Psi(I)$, and coinciding with $\Psi(I)$ for such intervals.

A necessary and sufficient condition that two distribution set-functions $\Phi_1$ and $\Phi_2$ be substantially equal is that $\int_{\Re_k} f(t)\, d\Phi_1 = \int_{\Re_k} f(t)\, d\Phi_2$ for any continuous function $f(t)$ which vanishes for all sufficiently large values of $|t|$.

In the one-dimensional case a distribution set-function $\Phi(e)$ generates a point-function $\psi(t)$, which may be defined, for instance, by setting $\psi(t) = \Phi(I_t) + C$, where $I_t$ is the infinite interval $-\infty < x \leqq t$, and $C$ is an arbitrary constant. This point function is increasing and bounded in $(-\infty, \infty)$ and is determined uniquely at all its points of continuity, up to an additional constant. Conversely, every point function which is increasing and bounded generates a distribution set-function which is determined substantially uniquely.

For this reason any bounded increasing point-function may be called simply a distribution function.

Two distribution functions are said to be substantially equal if they have the same points of continuity and if their values at common points of continuity differ only by a constant. A function $\psi(t)$ which is increasing and bounded in a finite closed interval $[a, b]$ can be extended over the interval $(-\infty, \infty)$ by setting $\psi(t) = \psi(a)$, $t < a$, $\psi(t) = \psi(b)$, $t > b$. It then becomes a distribution function. Two functions $\psi_1(t)$, $\psi_2(t)$ which are increasing and bounded over a finite closed interval $[a, b]$ are said to be substantially equal if they have the same interior points of continuity and if their values at these points, and also at the end-points $t = a$, $t = b$, differ by a constant. Analogous considerations hold, of course, in the general $k$-dimensional case.

For proofs we refer to [Bochner, 1; Haviland, 2].

**3. Theorems of Helly.** A sequence of additive functions of intervals $\{\Psi_n(I)\}$ is said to *converge substantially* to a function of intervals $\Psi(I)$ if $\lim_{n\to\infty} \Psi_n(I) = \Psi(I)$ for all (finite) intervals of continuity of $\Psi$.

"FIRST THEOREM OF HELLY". *Given a sequence $\{\Psi_n(I)\}$ of non-negative additive and uniformly bounded functions of intervals, then there exists a subsequence $\{\Psi_{n_i}(I)\}$ and a distribution function $\Phi$ to which this subsequence converges substantially. Furthermore, if the sequence $\{\Psi_n\}$ itself does not converge substantially to $\Phi$, then there exists another subsequence $\{\Psi_{n_i'}(I)\}$ converging substantially to another distribution function $\Phi'$ which is not substantially equal to $\Phi$.*

"SECOND THEOREM OF HELLY". *Given a sequence $\{\Psi_n(I)\}$ of non-negative additive and uniformly bounded functions of intervals, which converges substantially to a distribution function $\Phi$. Then*

$$\lim_n \int_{\Re_k} f(t)\, d\Psi_n = \int_{\Re_k} f(t)\, d\Phi$$

*for any function $f(t)$ continuous in $\Re_k$ and such that, as $I_N \uparrow \Re_k$, $\int_{I_N} f(t)\, d\Psi_n \to \int_{\Re_k} f(t)\, d\Psi_n$ uniformly in $n$.*

In the one-dimensional case this theorem may be easily restated in terms of sequences of uniformly bounded increasing point-functions, instead of functions of intervals. For the proof see [Bochner, 1].

**4. Extension theorem for non-negative functionals.** *Let $\mathfrak{M}$ be a linear manifold* of real-valued functions $x(t)$ defined on any abstract space $\Omega$, $t \in \Omega$. Let $\mathfrak{M}_0$ be a linear sub-manifold of $\mathfrak{M}$ and let $f_0(x)$ be a $(\Omega_0)$ non-negative additive and homogeneous functional defined on $\mathfrak{M}_0$, that is*

$$f_0(x_1 + x_2) = f_0(x_1) + f_0(x_2), \qquad x_1, x_2 \in \mathfrak{M}_0,$$

$$f_0(cx) = cf(x), \qquad x \in \mathfrak{M}_0,$$

$$f_0(x) \geqq 0, \quad \textit{whenever} \quad x(t) \geqq 0 \quad \textit{for all} \quad t \in \Omega_0 \subset \Omega.^{**}$$

*Let $\mathfrak{M}_0$ have the following property: (†) For every $y \in \mathfrak{M}$ there exist $x'$ and $x''$ of $\mathfrak{M}_0$ such that $x'(t) \leqq y(t) \leqq x''(t)$ on $\Omega_0$. Then the functional $f_0(x)$ can be extended to an additive, homogeneous and $(\Omega_0)$ non-negative functional $f(x)$ defined on the whole manifold $\mathfrak{M}$ so that $f(x)$ coincides with $f_0(x)$ on $\mathfrak{M}_0$.*

Assume $\mathfrak{M}_0 \subset \mathfrak{M}$ and $y_1 \in \mathfrak{M} - \mathfrak{M}_0$. Consider the linear manifold $\mathfrak{M}_1$ determined by $\mathfrak{M}_0$ and $y_1$. The elements $x_1$ of $\mathfrak{M}_1$ admit of a unique representation $x_1 = x_0 + uy_1$, where $x_0$ is any element of $\mathfrak{M}_0$ and $u$ is any real number. Introduce the functional $f(x_1)$ defined on $\mathfrak{M}_1$ by

$$f(x_1) = f(x_0 + uy_1) = f_0(x_0) + ur_1, \qquad r_1 = f(y_1).$$

It is clear that this functional is additive and homogeneous, and that it coincides with $f_0(x)$ when $x \in \mathfrak{M}_0$. It remains to determine $r_1$ so that $f$ will be $(\Omega_0)$ non-negative. Take any $x_0'$ and $x_0''$ such that $x_0' \leqq y_1 \leqq x_0''$ on $\Omega_0$. Then the condition

* That is, a set of functions $x(t)$ which contains $cx(t)$, $x(t) + y(t)$, whenever $x(t)$, $y(t)$ belong to the set, and $c$ is a real constant.

** $\Omega_0$ is any given set in $\Omega$; in particular, $\Omega_0$ may coincide with $\Omega$.

that $f_0(x)$ is $(\Omega_0)$ non-negative implies that $f_0(x_0') \leqq f_0(x_0'')$. Take $r_1$ to satisfy $\sup f_0(x') \leqq r_1 \leqq \inf f_0(x'')$, where sup and inf are taken over all $x' \leqq y_1$ and $x'' \geqq y_1$, respectively. Then $f$ will be $(\Omega_0)$ non-negative. For, if $x_0 + uy_1 \geqq 0$ on $\Omega_0$ and $u > 0$, then $y_1 \geqq -u^{-1}x_0$ and so $r_1 \geqq f(-u^{-1}x_0)$, i.e. $f(x_0) + ur_1 \geqq 0$; similarly, if $u < 0$, $-u^{-1}x_0 \geqq y_1$, and so $f(-u^{-1}x_0) \geqq r_1$, i.e. $f(x_0) + ur_1 \geqq 0$. Thus $f_0(x)$ is extended to the linear manifold $\mathfrak{M}_1$. The extension to the whole linear manifold $\mathfrak{M}$ can now be performed by the method of transfinite induction.

The proof above proceeds along the same lines as a proof by Kantorovich [1]. Compare also Haviland [4, 5]. The condition (†) was suggested by a result of Krein and Šmulian [1]; it was omitted in the first printing. The proof in the first printing tacitly assumed that $\inf f_0(x_0)$ for $x_0 - y_1 \geqq 0$ on $\Omega_0$ is not $-\infty$. Nothing in the original statement guarantees this, and if it is not true the extended "functional" would not be finite-valued. In the application on the next page to the proof of Theorem 1.1, condition (1.3) ensures the validity of (†).

**5. Stieltjes inversion formula.** Let $\psi(t)$ be any function of bounded variation on $(-\infty, \infty)$. The integral $I(z) \equiv I(z; \psi) = \int_{-\infty}^{\infty} \frac{d\psi(t)}{z - t}$ is an analytic function of $z$ in the upper and in the lower half-planes, its values being conjugate at two conjugate points. The function $\psi(t)$ can be expressed in terms of $I(z)$ by the following formula:

$$\tfrac{1}{2}[\psi(t_1 + 0) + \psi(t_1 - 0)] - \tfrac{1}{2}[\psi(t_0 + 0) + \psi(t_0 - 0)]$$
$$= \lim_{\epsilon \to +0} -\frac{1}{2\pi i} \int_{t_0}^{t_1} [I(t + i\epsilon) - I(t - i\epsilon)]\, dt.$$

(Cf. Stone, [1]). Thus, $\psi(t)$ is substantially uniquely determined by $I(z; \psi)$.

## CHAPTER I

# A GENERAL THEORY OF THE PROBLEM OF MOMENTS.

**1.** Let $\mathfrak{R}$ be a $k$-dimensional Euclidean space. Let there be given an infinite multiple sequence of real constants

$$\mu_{j_1 j_2 \cdots j_k}; \qquad j_1, j_2, \cdots, j_k = 0, 1, 2, \cdots .$$

We are interested in finding necessary and sufficient conditions that there shall exist a $k$-dimensional distribution function $\Phi$ whose spectrum $\mathfrak{S}(\Phi)$ is to be contained in a closed set $\mathfrak{S}_0$, given in advance, and which is a solution of the "problem of moments" [Haviland, 4, 5]

$$(1.1) \qquad \mu_{j_1 \cdots j_k} = \int_{\mathfrak{R}} t_1^{j_1} \cdots t_k^{j_k} \, d\Phi, \qquad j_1, j_2, \cdots, j_k = 0, 1, 2, \cdots$$

To abbreviate we call this problem simply the ($\mathfrak{S}_0$) moment problem. We say that the moment problem is determined if its solution is substantially unique; otherwise we call it indeterminate.

To simplify we shall discuss only the two-dimensional case, $k = 2$. There is no difficulty in extending the results to the case of any number of dimensions.

Let $P(u, v)$ be any polynomial in $u$, $v$ in $\mathfrak{R}$,

$$P(u, v) = \sum_{i,j} x_i y_j u^i v^j,$$

where $x_i$, $y_j$ are real- or complex-valued constants. Introduce the functional $\mu(P)$ defined by

$$\mu(P) \equiv \sum_{i,j} \mu_{ij} x_i y_j .$$

In particular,

$$\mu(u^i v^j) = \mu_{ij} .$$

THEOREM 1.1. *A necessary and sufficient condition that the* ($\mathfrak{S}_0$) *moment problem defined by the sequence of moments* $\{\mu_{ij}\}$ *shall have a solution is that the functional* $\mu(P)$ *be* ($\mathfrak{S}_0$) *non-negative, that is*

$$(1.2) \qquad \mu(P) \geqq 0, \quad \textit{whenever} \quad P(u, v) \geqq 0 \quad \textit{on} \quad \mathfrak{S}_0 .$$

This theorem is an immediate application of the theorem on the extension of non-negative functionals (Introduction, **4**). Let $\mathfrak{M}$ be the linear manifold of all single-valued functions $y = y(u, v)$ which admit of an estimate

$$(1.3) \qquad | y(u, v)| \leqq A(u^{2r} + v^{2r}) + B,$$

where $A$, $B$ are non-negative constants and $r$ is a non-negative integer. Let $\mathfrak{M}_0$ be the linear sub-manifold of $\mathfrak{M}$, consisting of all polynomials $P$. It is clear that all functions $\eta = A(u^{2r} + v^{2r}) + B$ are in $\mathfrak{M}_0$.

Now if our $(\mathfrak{S}_0)$ moment-problem has a solution $\Phi$, then whenever $P \geqq 0$ on $\mathfrak{S}_0$, we obviously have

$$\mu(P) = \int_{\mathfrak{R}} P(u, v)\, d\Phi = \int_{\mathfrak{S}_0} P(u, v)\, d\Phi \geqq 0.$$

Thus the condition (1.2) of Theorem 1.1 is necessary. To prove its sufficiency, suppose (1.2) is satisfied. Then $\mu(P)$ appears as a homogeneous additive $(\mathfrak{S}_0)$ non-negative functional defined on $\mathfrak{M}_0$. By Introduction, **4**, this functional can be extended over the whole manifold $\mathfrak{M}$, with preservation of all these properties. In particular, we may define $\mu(y_I)$, where $y_I$ is the characteristic function of any two-dimensional interval $I$, since clearly $y_I \in \mathfrak{M}$. Thus we obtain a function $\psi(I) = \mu(y_I)$ of intervals, which possesses the properties

(i) $$\psi(I) \geqq 0,$$

since $y_I \geqq 0$ and $\mu$ is non-negative;

(ii) whenever $I = \sum_{i=1}^{n} I_i$, $I_i I_j = 0$ for $i \neq j$, then

$$\psi(I) = \sum_{i=1}^{n} \psi(I_i),$$

since $\mu$ is additive and $y_I = \sum_{i=1}^{n} y_{I_i}$;

(iii) $\psi(I)$ is bounded,

since

$$\psi(I) \leqq \psi(\mathfrak{R}) = \mu(y_{\mathfrak{R}}) = \mu(1) = \mu_{00}.$$

The conditions of Introduction, **2**, are thus satisfied, and we can construct the associated distribution function $\Phi$ which is substantially equal to $\psi(I)$. This function $\Phi$ is a solution of our $(\mathfrak{S}_0)$ moment problem. To prove this we have only to establish that

$$\mathfrak{S}(\Phi) \leqq \mathfrak{S}_0, \tag{1.4}$$

$$\int_{\mathfrak{R}} u^i v^j\, d\Phi = \mu_{ij}, \qquad i, j = 0, 1, 2, \cdots. \tag{1.5}$$

To prove (1.4) it suffices to show that $(u_0, v_0) \in \mathfrak{R} - \mathfrak{S}_0$ implies $(u_0, v_0) \in \mathfrak{R} - \mathfrak{S}(\Phi)$. Let $(u_0, v_0) \in \mathfrak{R} - \mathfrak{S}_0$ and let $I_0 < \mathfrak{R} - \mathfrak{S}_0$ be a common interval of continuity of $\Phi$ and $\psi$ containing $(u_0, v_0)$ in its interior. Since $y_{I_0} = 0$ on $\mathfrak{S}_0$ we may write $y_{I_0} \leqq 0$, $y_{I_0} \geqq 0$ on $\mathfrak{S}_0$, whence $\mu(y_{I_0}) \leqq 0$, $\mu(y_{I_0}) \geqq 0$, and $\mu(y_{I_0}) = 0$. Thus $\psi(I_0) = \mu(y_{I_0}) = 0$, which implies $\Phi(I_0) = 0$; hence $(u_0, v_0) \in \mathfrak{R} - \mathfrak{S}(\Phi)$.

It remains to prove (1.5). Let $\epsilon$, $\epsilon_1$ be two given positive numbers. Let $I_0$ be again a common interval of continuity of $\Phi$ and $\psi$, so large that for a suitable choice of $r$

$$|u^i v^j| < \epsilon(u^{2r} + v^{2r}) \quad \text{on} \quad \mathfrak{R} - I_0 .$$

The integers $i, j$, and $r$ will be now fixed. Let $I_1, I_2, \cdots, I_n$ be a finite sequence of common intervals of continuity of $\Phi$ and $\psi$, disjoint and such that

$$I_0 = I_1 + I_2 + \cdots + I_n ,$$

while the oscillation of $u^i v^j$ on each $\bar{I}_\nu$, $\nu = 1, 2, 3, \cdots, n$ is less than $\epsilon_1$. In each $I_\nu$ select a point $(u_\nu, v_\nu)$ and introduce a simple function

$$y_0(u, v) = \begin{cases} u_\nu^i v_\nu^j & \text{on} \quad I_\nu , \qquad \nu = 1, 2, \cdots, n, \\ 0 \text{ elsewhere.} \end{cases}$$

It is clear that

$$y_0(u, v) = \sum_{\nu=0}^{n} u_\nu^i v_\nu^j y_{I_\nu} .$$

Since

$$y_0(u, v) - \epsilon_1 < u^i v^j < y_0(u, v) + \epsilon_1 \quad \text{on} \quad I_\nu , \qquad \nu = 1, 2, \cdots, n,$$

while

$$-\epsilon(u^{2r} + v^{2r}) < u^i v^j < \epsilon(u^{2r} + v^{2r}) \quad \text{on} \quad \mathfrak{R} - I_0 ,$$

we have everywhere in $\mathfrak{R}$

$$y_0(u, v) - \epsilon(u^{2r} + v^{2r}) - \epsilon_1 < u^i v^j < y_0(u, v) + \epsilon(u^{2r} + v^{2r}) + \epsilon_1 .$$

In view of the $(\mathfrak{S}_0)$ non-negativeness of the functional $\mu$, we have

$$\mu[y_0(u, v) - \epsilon(u^{2r} + v^{2r}) - \epsilon_1] \leqq \mu(u^i v^j) \leqq \mu[y_0(u, v) + \epsilon(u^{2r} + v^{2r}) + \epsilon_1],$$

$$\mu(y_0) - \epsilon(\mu_{2r,0} + \mu_{0,2r}) - \epsilon_1\mu_{00} \leqq \mu_{ij} \leqq \mu(y_0) + \epsilon(\mu_{2r,0} + \mu_{0,2r}) + \epsilon_1\mu_{00} .$$

But

$$\mu(y_0) = \sum_{\nu=1}^{n} u_\nu^i v_\nu^j \mu(y_{I_\nu}) = \sum_{\nu=1}^{n} u_\nu^i v_\nu^j \psi(I_\nu) = \sum_{\nu=1}^{n} u_\nu^i v_\nu^j \Phi(I_\nu).$$

If we allow here $\epsilon_1 \to 0$ and $\max |I_\nu| \to 0$, we see that

$$\int_{I_0} u^i v^j \, d\Phi - \epsilon(\mu_{2r,0} + \mu_{0,2r}) \leqq \mu_{ij} \leqq \int_{I_0} u^i v^j \, d\Phi + \epsilon(\mu_{2r,0} + \mu_{0,2r}).$$

On allowing here $\epsilon \to 0$ (and $I_0 \to \mathfrak{R}$) it readily follows that $u^i v^j$ is absolutely integrable over $\mathfrak{R}$ with respect to $\Phi$ and that

$$\mu_{ij} = \int_{\mathfrak{R}} u^i v^j \, d\Phi, \qquad i, j = 0, 1, 2, \cdots .$$

**2.** Theorem 1.1 can be readily applied to derive necessary and sufficient conditions for the existence of solutions of various specialized moment problems characterized by a special choice of $\mathfrak{S}_0$.

(a) HAMBURGER MOMENT PROBLEM. Here $\mathfrak{S}_0$ coincides with the axis of reals. Hence $\mu_{ij} = 0$ for $j \geqq 1$, so that we have a simple sequence of moments

$$\mu_n \equiv \mu_{n0}, \qquad n = 0, 1, 2, \cdots,$$

and the problem reduces to that of determining a one-dimensional distribution function $\psi(u)$ such that

$$\mu_n = \int_{-\infty}^{\infty} u^n \, d\psi(u), \qquad n = 0, 1, 2, \cdots. \tag{1.6}$$

Thus it suffices to consider polynomials and functions of $u$ alone, and to define the functional $\mu$ by

$$\mu(P_n) \equiv \sum_{j=0}^{n} \mu_j x_j, \qquad P_n(u) = \sum_{j=0}^{n} x_j u^j. \tag{1.7}$$

Theorem 1.1 states now that a necessary and sufficient condition for the existence of a solution of (1.6) is that $\mu(P) \geqq 0$, whenever $P(u) \geqq 0$ for all real values of $u$. If we take for $P(u)$ the particular polynomial $P(u) = (x_0 + x_1 u + \cdots + x_n u^n)^2$, $x_i$ real, we have

$$\mu(P) = \sum_{i,j=0}^{n} \mu_{i+j} x_i x_j \equiv Q_n(x) \tag{1.8}$$

(Hankel quadratic form). Thus *a necessary condition for the existence of a solution of* (1.6) *is that the quadratic forms* $Q_n(x)$, $n = 0, 1, 2, \cdots$, *be non-negative. This condition is also sufficient.*

In fact [Pólya und Szegö, 1, Vol. II, p. 82] any polynomial $P(u) \geqq 0$ for all real $u$ can be represented by

$$P(u) = p_1(u)^2 + p_2(u)^2,$$

where $p_1(u)$, $p_2(u)$ are polynomials with real coefficients, whence

$$u(P) = \mu(p_1^2) + \mu(p_2^2) \geqq 0,$$

if $Q_n(x) \geqq 0, n = 0, 1, 2, \cdots$.

Let $\psi(u)$ be a solution of (1.6). Since

$$Q_n(x) = \mu(p_n^2) = \int_{-\infty}^{\infty} p_n^2(t) \, d\psi,$$

it is clear that if $\mathfrak{S}(\psi)$ is not reducible to a finite set of points, we always have

$$Q_n(x) = \sum_{i,j=0}^{n} \mu_{i+j} x_i x_j > 0, \qquad n = 0, 1, 2, \cdots, \tag{1.9}$$

provided not all $x_0, x_1, \cdots, x_n$ are zero, which will be assumed in what follows. From the theory of quadratic forms it is well known that conditions (1.9) are equivalent to

$$\Delta_n = |\mu_{i+j}|_{i,j=0}^{n} > 0, \qquad n = 0, 1, 2, \cdots.$$

On the other hand, if there exists a solution $\psi(u)$ whose spectrum consists precisely of $(k + 1)$ distinct points (we shall see later [**3**, Corollary 1.1] that the moment problem is then determined), $t_1, t_2, \cdots, t_{k+1}$, then it is readily seen that for each $n \geqq k + 1$, $Q_n(x) = 0$ for a suitable choice of $x_0, x_1, \cdots, x_n$, which implies $\Delta_n = 0$, $n = k + 1, k + 2, \cdots$, while $\Delta_0 > 0, \cdots, \Delta_k > 0$. It can be proved, conversely, that if these conditions are satisfied, then there exists a uniquely determined solution of the moment problem with the property mentioned above [Fischer, 1; Achyeser and Krein, 1, 6].

All these results can be stated in

THEOREM 1.2. *In order that a Hamburger moment problem*

$$\mu_n = \int_{-\infty}^{\infty} t^n \, d\psi, \qquad n = 0, 1, 2, \cdots, \tag{1.6}$$

*shall have a solution it is necessary that*

$$\Delta_n = | \mu_{i+j} |_{i,j=0}^{n} \geqq 0, \qquad n = 0, 1, 2, \cdots. \tag{1.10}$$

*In order that there exist a solution whose spectrum is not reducible to a finite set of points it is necessary and sufficient that*

$$\Delta_n > 0, \qquad n = 0, 1, 2, \cdots. \tag{1.11}$$

*In order that there exist a solution whose spectrum consists of precisely $(k + 1)$ distinct points it is necessary and sufficient that*

$$\Delta_0 > 0, \cdots, \Delta_k > 0, \qquad \Delta_{k+1} = \Delta_{k+2} = \cdots = 0. \tag{1.12}$$

*The moment problem is determined in this case.*

(b) STIELTJES MOMENT PROBLEM. In this case $\mathfrak{S}_0$ coincides with the positive part of the axis of reals, $u \geqq 0$. As in the preceding case, we have to consider only moments $\mu_n = \mu_{n0}$, and only polynomials and functions of a single variable. The moment problem reduces to

$$\mu_n = \int_0^{\infty} t^n d\psi, \qquad n = 0, 1, 2, \cdots,$$

and a necessary and sufficient condition for the existence of a solution is that

$$\mu(P) = \sum_{j=0}^{n} \mu_j x_j \geqq 0,$$

whenever

$$P(u) = x_0 + x_1 u + \cdots + x_n u^n \geqq 0 \quad \text{for} \quad u \geqq 0.$$

An application of this condition to the two special polynomials $(x_0 + x_1 u + \cdots + x_n u^n)^2$, $u(x_0 + x_1 u + \cdots + x_n u^n)^2$ yields at once

$$\left.\begin{aligned} Q_n(x) &\equiv \sum_{i,j=0}^{n} \mu_{i+j} x_i x_j \geqq 0, \\ Q_n^{(1)}(x) &\equiv \sum_{i,j=0}^{n} \mu_{i+j+1} x_i x_j \geqq 0, \end{aligned}\right\} \qquad n = 0, 1, 2, \cdots,$$

which is equivalent to

$$\Delta_n = |\mu_{i+j}|_{i,j=0}^{n} \geqq 0, \qquad \Delta_n^{(1)} = |\mu_{i+j+1}|_{i,j=0}^{n} \geqq 0, \qquad n = 0, 1, 2, \cdots .$$

As in the preceding case, we derive

THEOREM 1.3. *A necessary condition for the existence of a solution of the Stieltjes moment problem*

$$(1.13) \qquad \mu_n = \int_0^{\infty} t^n \, d\psi, \qquad n = 0, 1, 2, \cdots,$$

*is that*

$$(1.14) \quad \Delta_n = |\mu_{i+j}|_{i,j=0}^{n} \geqq 0, \quad \Delta_n^{(1)} = |\mu_{i+j+1}|_{i,j=0}^{n} \geqq 0, \quad n = 0, 1, 2, \cdots .$$

*In order that there exist a solution whose spectrum is not reducible to a finite set of points it is necessary and sufficient that*

$$(1.15) \qquad \Delta_n > 0, \qquad \Delta_n^{(1)} > 0, \qquad n = 0, 1, 2, \cdots .$$

*In order that there exist a solution whose spectrum consists of precisely* $(k + 1)$ *points distinct from* $t = 0$ *it is necessary and sufficient that*

$$(1.16) \qquad \begin{cases} \Delta_0 > 0, \cdots, \Delta_k > 0, \Delta_{k+1} = \Delta_{k+2} = \cdots = 0, \\ \Delta_0^{(1)} > 0, \cdots, \Delta_k^{(1)} > 0, \Delta_{k+1}^{(1)} = \Delta_{k+2}^{(1)} = \cdots = 0, \end{cases}$$

*while in order that there exist a solution whose spectrum consists of* $(k + 1)$ *points, one of them being at* $t = 0$, *it is necessary and sufficient that*

$$(1.17) \qquad \begin{cases} \Delta_0 > 0, \cdots, \Delta_k > 0, \Delta_{k+1} = \Delta_{k+2} = \cdots = 0, \\ \Delta_0^{(1)} > 0, \cdots, \Delta_{k-1}^{(1)} > 0, \Delta_k^{(1)} = \Delta_{k+1}^{(1)} = \cdots = 0. \end{cases}$$

*In the last two cases the problem is determined.**

(c) TRIGONOMETRIC MOMENT PROBLEM. Here the set $\mathfrak{S}_0$ is the circumference of a circle which, without loss of generality, can be taken as the unit circle. Since in this case $u = \cos\theta$, $v = \sin\theta$, and every polynomial of degree $n$ in $\cos\theta$, $\sin\theta$ can be written as $P_n(\theta) = \sum_{k=-n}^{n} x_k e^{ik\theta}$, and conversely, where $x_k$ are complex numbers, we again may replace the double sequence of moments $\mu_{ij}$ by a simple sequence $\{\mu_n\}$, $n = 0, \pm 1, \pm 2, \cdots$, $\mu_{-n} = \bar{\mu}_n$, and introduce the "trigonometric moment problem",

$$\mu_n = \int_0^{2\pi} e^{in\theta} \, d\Phi, \qquad n = 0, \pm 1, \pm 2, \cdots; \qquad \mu_{-n} = \bar{\mu}_n .$$

Theorem 1.1 shows that a necessary and sufficient condition for the existence of a solution of the trigonometric moment problem is that

$$P_n(\theta) = \sum_{k=-n}^{n} x_k e^{ik\theta} \geqq 0 \quad \text{for all} \quad \theta$$

* Here we have used the fact that a polynomial $P(u) \geqq 0$ for $u \geqq 0$ can be written in the form $p_1(u)^2 + up_2(u)^2$.

implies

$$\mu(P_n) = \sum_{k=-n}^{n} \mu_k x_k \geqq 0, \qquad n = 0, 1, 2, \cdots .$$

Since, on the other hand, a general representation for a non-negative trigonometric polynomial $P_n(\theta)$ is [Pólya und Szegö, 1, Vol. II, p. 82]

$$P_n(\theta) = | h_n(z) |^2, \qquad h_n(z) = \sum_{j=0}^{n} x_j z^j, \qquad z = e^{i\theta},$$

the expression $\mu(P_n) = \mu(| h_n(z) |^2) = \mu\left(\sum_{j,l=0}^{n} x_j e^{ij\theta} \bar{x}_l e^{-il\theta}\right)$ reduces to the Hermitian form

$$\sum_{j,l=0}^{n} \mu_{j-l}\, x_j \bar{x}_l$$

(Toeplitz form), and we obtain

THEOREM 1.4. *A necessary condition that the trigonometric moment problem*

$$\mu_n = \int_0^{2\pi} e^{in\theta}\, d\Phi, \qquad n = 0, \pm 1, \pm 2, \cdots; \qquad \mu_{-n} = \bar{\mu}_n ,$$

*have a solution is that all Toeplitz forms*

$$\sum_{j,l=0}^{n} \mu_{j-l} x_j \bar{x}_l \geqq 0, \qquad n = 0, 1, 2, \cdots ,$$

*or else that all determinants*

$$\delta_n = | \mu_{i-j} |_{i,j=0}^{n} \geqq 0, \qquad n = 0, 1, 2, \cdots .$$

*For the existence of a solution whose spectrum is not reducible to a finite set of points it is necessary and sufficient that*

$$\delta_n > 0, \qquad n = 0, 1, 2, \cdots .$$

We shall see later that the trigonometric moment problem is always determined.

Using an analogous argument it can be shown [Achyeser and Krein, 6] that *a necessary and sufficient condition for the existence of a solution of the moment problem*

$$\mu_n = \int_{-\tau}^{\tau} e^{in\theta}\, d\Phi, \qquad n = 0, \pm 1, \pm 2, \cdots; \qquad \mu_{-n} = \bar{\mu}_n , \qquad 0 < \tau < \pi,$$

*is that $\mu_0 \geqq 0$, and all the forms*

$$\sum_{j,l=0}^{n} \mu_{j-l} x_j \bar{x}_l , \qquad \sum_{j,l=0}^{n-1} (\mu_{j-l+1} - 2 \cos\tau\, \mu_{j-l} + \mu_{j-l-1}) x_j \bar{x}_l$$

*be non-negative.* It is easy to show that this criterion reduces to that of Theorem 1.4 when $\tau = \pi$.

(d) THE HAUSDORFF ONE-DIMENSIONAL MOMENT PROBLEM. The set $\mathfrak{S}_0$ reduces here to the closed interval [0, 1] of the axis of reals. Using the notation

$$\mu_n = \int_0^1 t^n \, d\Phi, \qquad n = 0, 1, 2, \cdots, \tag{1.18}$$

$$\mu(P_n) = \sum_{j=0}^{n} \mu_j x_j, \qquad P_n(u) = \sum_{j=0}^{n} x_j u^j, \tag{1.19}$$

we conclude that a necessary and sufficient condition for the existence of a solution of (1.18) is that

$$P_n(u) \geqq 0 \quad \text{on} \quad [0, 1] \quad \text{implies} \quad \mu(P_n) \geqq 0. \tag{1.20}$$

To transform this condition we have to discuss the representation of polynomials non-negative on [0, 1]. Let $f(t)$ be any function single-valued and finite on [0, 1]. The polynomial of degree $n$,

$$B_n(t; f) = \sum_{\nu=0}^{n} f\left(\frac{\nu}{n}\right)\binom{n}{\nu} t^\nu (1 - t)^{n-\nu}$$

is called the Bernstein polynomial of degree $n$ associated with $f(t)$. [S. Bernstein, 1].

An important property of Bernstein polynomials is expressed by the following statement.

If $P_m(t)$ is a fixed polynomial of degree $m$, then

$$B_n(t; P_m) = P_m(t) + \sum_{s=1}^{m-1} \frac{p_{ms}(t)}{n^s},$$

where the polynomials $p_{ms}(t)$ do not depend on $n$ (and are $\equiv 0$ in case $P_m(t) =$ const.).

It is easy to show now that the condition (1.20) is equivalent to

$$\mu(t^\nu(1 - t)^{n-\nu}) \geqq 0, \qquad \nu = 0, 1, \cdots, n; \qquad n = 0, 1, \cdots. \tag{1.21}$$

Indeed, $t^\nu(1 - t)^{n-\nu}$ clearly is $\geqq 0$ on (0, 1) so that (1.20) implies (1.21). Conversely, assume (1.21) to be satisfied and let $P_m(t)$ be any polynomial of degree $m$, non-negative on (0, 1). Construct

$$B_n(t; P_m) = P_m(t) + \sum_{s=1}^{m-1} \frac{p_{ms}(t)}{n^s}.$$

Then we have

$$\mu(B_n) = \mu(P_m) + \sum_{s=1}^{m-1} \mu(p_{ms})n^{-s} = \mu(P_m) + O(1/n).$$

On the other hand, since $B_n$ is a linear combination, with positive coefficients, of expressions of the form $t^\nu(1 - t)^{n-\nu}$, we see that (1.21) implies $\mu(B_n) \geqq 0$. On allowing here $n \to \infty$ it follows that $\mu(P_m) \geqq 0$.*

* A shorter proof could be based on a known representation of polynomials non-negative on (0, 1) (Pólya und Szegö, 1, Vol. II, p. 82). However, the proof in the text (due to Hildebrandt and Schoenberg [1]) can be extended to any number of dimensions without any essential modifications.

Introducing the differences

$$\Delta^0\mu_\nu = \mu_\nu,$$
$$\Delta^1\mu_\nu = \mu_\nu - \mu_{\nu+1},$$
$$\cdots\cdots\cdots\cdots\cdots\cdots\cdots\cdots$$
$$\Delta^k\mu_\nu = \mu_\nu - \binom{k}{1}\mu_{\nu+1} + \binom{k}{2}\mu_{\nu+2} + \cdots + (-1)^k\mu_{\nu+k} = \mu(t^\nu(1-t)^k),$$

we may state (1.21) in an equivalent form:

THEOREM 1.5. *A necessary and sufficient condition that the one-dimensional Hausdorff moment problem*

$$(1.18) \qquad \mu_n = \int_0^1 t^n\, d\Phi, \qquad n = 0, 1, 2, \cdots,$$

*shall have a solution is that all differences*

$$\Delta^k\mu_\nu \geqq 0, \qquad k, \nu = 0, 1, 2, \cdots.$$

(e) THE HAUSDORFF TWO-DIMENSIONAL MOMENT PROBLEM. [Haviland, 4, 5]. Here the set $\mathfrak{S}_0$ is a rectangle in the $(u, v)$-plane, which, without loss of generality, may be taken as the unit square $0 \leqq u \leqq 1, 0 \leqq v \leqq 1$. By introducing the Bernstein polynomials in two variables associated with a given function $f(u, v)$,

$$B_{n,m}(u, v; f) \equiv \sum_{i=0}^{n}\sum_{j=0}^{m} f\left(\frac{i}{n}, \frac{j}{m}\right)\binom{n}{i}\binom{m}{j} u^i(1-u)^{n-i}v^j(1-v)^{m-j}$$

and the double differences of moments,

$$\Delta_1^n\,\Delta_2^m\,\mu_{ij} = \sum_{r=0}^{n}\sum_{s=0}^{m}\binom{n}{r}\binom{m}{s}(-1)^{r+s}\mu_{i+r,j+s},$$

we can immediately extend the preceding considerations to prove

THEOREM 1.6. *A necessary and sufficient condition that the two-dimensional Hausdorff moment-problem*

$$\mu_{ij} = \int_0^1\int_0^1 u^iv^j\, d\Phi, \qquad i, j = 0, 1, 2, \cdots,$$

*shall have a solution is that all differences*

$$\Delta_1^n\Delta_2^m\mu_{ij} \geqq 0; \qquad n, m = 0, 1, 2, \cdots; \qquad i, j = 0, 1, 2, \cdots.$$

As we shall see later one- and two-dimensional Hausdorff moment-problems are always determined.

**3.** We now turn to a discussion of the uniqueness of the solution of the $(\mathfrak{S}_0)$ moment-problem. We have agreed to say that the problem is determined if the difference of any two solutions is substantially equal to a constant. Again let $\mathfrak{M}$ be the manifold of single-valued functions $y(u, v)$ which admit of the estimate (1.3). Let $\mathfrak{M}_0$ be the sub-manifold of $\mathfrak{M}$ consisting of all polynomials, and let

$\mathfrak{M}_C$ be the sub-manifold of continuous function of $\mathfrak{M}$. We also admit complex-valued functions as members of $\mathfrak{M}$. In this case we denote by $\mathfrak{M}_C^c$ the sub-manifold of all complex-valued continuous functions of $\mathfrak{M}$. We reserve letters $p$, $P$, with various subscripts, to designate members of $\mathfrak{M}_0$.

Let $y$ be an arbitrary element of $\mathfrak{M}$. Introduce two functionals

$$\underline{\mu}(y) = \sup_{p \leqq y \text{ on } \mathfrak{S}_0} \mu(p), \qquad \bar{\mu}(y) = \inf_{p \geqq y \text{ on } \mathfrak{S}_0} \mu(p).$$

It is clear that these functionals are quasi-homogeneous:

$$\underline{\mu}(cy) = c\underline{\mu}(y), \qquad \bar{\mu}(cy) = c\bar{\mu}(y), \qquad c \geqq 0,$$

while

$$\underline{\mu}(y_1 + y_2) \geqq \underline{\mu}(y_1) + \underline{\mu}(y_2), \qquad \bar{\mu}(y_1 + y_2) \leqq \bar{\mu}(y_1) + \bar{\mu}(y_2).$$

It is also obvious that

$$\underline{\mu}(y) \leqq \bar{\mu}(y).$$

THEOREM 1.7. *Assume that the* ($\mathfrak{S}_0$) *moment-problem has a solution* $\Phi$. *Let* $y_0$ *be a fixed function of* $\mathfrak{M}_C$. *The range of values assumed by the integral*

$$\int_{\mathfrak{R}} y_0(u, v)\, d\Phi, \tag{1.22}$$

*when* $\Phi$ *ranges over all possible solutions of the* ($\mathfrak{S}_0$) *moment-problem, is the closed interval* $[\underline{\mu}(y_0), \bar{\mu}(y_0)]$. *Thus a necessary and sufficient condition that the* ($\mathfrak{S}_0$) *moment problem be determined is that*

$$\underline{\mu}(y_0) = \bar{\mu}(y_0) \quad \textit{for all} \quad y_0 \in \mathfrak{M}_C .$$

The integral (1.22) is an extension of the non-negative functional $\mu(P)$ to the linear manifold $\mathfrak{M}_C$. Thus

$$\underline{\mu}(y_0) \leqq \int_{\mathfrak{R}} y_0\, d\Phi \leqq \bar{\mu}(y_0).$$

On the other hand, let $l$ be an arbitrary number such that

$$\underline{\mu}(y_0) \leqq l \leqq \bar{\mu}(y_0).$$

In applying the extension theorem (Introduction, 4), if $y_0$ is not a polynomial, we may start our extension with $y_0$, defining

$$\mu(y_0) = l.$$

We have only to repeat the arguments given on p. 3, replacing $u^i v^j$ by $y_0(u, v)$, to prove the existence of a solution of the ($\mathfrak{S}_0$) moment problem such that

$$\int_{\mathfrak{R}} y_0(u, v)\, d\Phi = \mu(y_0) = l.$$

The necessity of the condition of Theorem 1.7 for the moment problem to be determined is obvious. If it is satisfied, then if $\Phi_1$, $\Phi_2$ are any two solutions, we must have $\int_{\mathfrak{R}} y_0\, d\Phi_1 = \int_{\mathfrak{R}} y_0\, d\Phi_2$ for any function $y_0 \in \mathfrak{M}_C$, in particular, for any continuous function vanishing outside a sufficiently large sphere. By Introduction **2**, it follows that $\Phi_1$ and $\Phi_2$ are substantially equal.

COROLLARY 1.1. *If a k-dimensional moment problem has a solution whose spectrum is a bounded set, then the moment problem is determined.*

Indeed, any function $y_0$ of $\mathfrak{M}_C$ can be approximated by two polynomials, $P_{1\epsilon}$, $P_{2\epsilon}$ so that in a fixed closed sphere,

$$P_{1\epsilon} \leqq y_0 \leqq P_{2\epsilon}, \qquad P_{2\epsilon} - P_{1\epsilon} < \epsilon,$$

where $\epsilon$ is an arbitrarily given positive number. If we take this sphere so large that it contains the spectrum of the solution $\Phi_0$ of the moment problem in question, and if we allow $\epsilon \to 0$, then it is readily seen that

$$\underline{\mu}(y_0) = \bar{\mu}(y_0).$$

Since $y_0$ is an arbitrary element of $\mathfrak{M}_C$, it follows that our moment problem is determined.

REMARK. In the case of the Hamburger or Stieltjes moment problem let

$$\underline{\mu}^{(n)}(y_0) = \sup_{p \leqq y_0 \text{ on } \mathfrak{S}_0} \mu(p), \qquad \bar{\mu}^{(n)}(y_0) = \inf_{p \geqq y_0 \text{ on } \mathfrak{S}_0} \mu(p),$$

where sup and inf are taken only over the set of polynomials $p$ of degree not greater than $2n$, and where $\underline{\mu}^{(n)}(y_0)$, $\bar{\mu}^{(n)}(y_0)$ are to be replaced by $-\infty$, $+\infty$ in case there is no polynomial of degree not greater than $2n$ which is respectively $\leqq y_0$ or $\geqq y_0$ on $\mathfrak{S}_0$. For sufficiently large values of $n$ both numbers $\underline{\mu}^{(n)}(y_0)$, $\bar{\mu}^{(n)}(y_0)$ are finite and

$$\underline{\mu}^{(n)}(y_0) \uparrow \underline{\mu}(y_0), \qquad \bar{\mu}^{(n)}(y_0) \downarrow \bar{\mu}(y_0), \qquad \text{as } n \uparrow \infty.$$

An analogous remark, of course, could be made for more general two- or many-dimensional moment-problems.

Theorem 1.7 can be readily extended to the case of a complex-valued function $y_0 \in \mathfrak{M}_C^c$.

THEOREM 1.8. *Let*

$$y_0(u, v) = y_0'(u, v) + iy_0''(u, v)$$

*be any given element of the manifold $\mathfrak{M}_C^c$. Let $\mathfrak{D}(y_0)$ be the range of values of the integral $\int_{\mathfrak{R}} y_0\, d\Phi$ when $\Phi$ ranges over all solutions of the $(\mathfrak{S}_0)$ moment-problem. Then $\mathfrak{D}(y_0)$ is a bounded convex and closed set. More precisely, if for any angle $\theta$, $0 \leqq \theta < 2\pi$, we define*

$$\underline{\mu}(\theta) = \underline{\mu}(y_0' \cos \theta + y_0'' \sin \theta), \qquad \bar{\mu}(\theta) = \bar{\mu}(y_0' \cos \theta + y_0'' \sin \theta)$$

*and for a given $\theta_0$ construct the closed strip in the $(X, Y)$ plane,*

$$(S_{\theta_0}) \qquad \underline{\mu}(\theta_0) \leqq X \cos \theta_0 + Y \sin \theta_0 \leqq \bar{\mu}(\theta_0),$$

*then $\mathfrak{D}(y_0)$ is the common part of all strips $S_{\theta_0}$, $0 \leqq \theta_0 < 2\pi$.*

That $\mathfrak{D}(y_0)$ is bounded is obvious. That it is convex follows from the fact that if $\Phi_1$, $\Phi_2$ are any two solutions of a $(\mathfrak{S}_0)$ moment problem, then $a\Phi_1 + (1 - a)\Phi_2$, $0 \leqq a \leqq 1$, is also a solution. Finally, that $\mathfrak{D}(y_0)$ is closed is readily proved on the basis of theorems of Helly and of the estimate (1.3).

We now pass to the second part of Theorem 1.8. Let $Z = X + iY$ be a point of the set $\mathfrak{D}(y_0)$ and let $\Phi$ be a corresponding solution of the $(\mathfrak{S}_0)$ moment-problem. Thus

$$Z = X + iY = \int_{\mathfrak{R}} y_0' \, d\Phi + i \int_{\mathfrak{R}} y_0'' \, d\Phi,$$

and

$$X \cos \theta + Y \sin \theta = \int_{\mathfrak{R}} (y_0' \cos \theta + y_0'' \sin \theta) \, d\Phi$$

must lie between $\underline{\mu}(\theta)$ and $\bar{\mu}(\theta)$. Since this holds for all $\theta$, we conclude that $\mathfrak{D}(y_0) \subseteq \mathfrak{D}$, where $\mathfrak{D}$ is the common part of all strips $S_\theta$ in question. It remains to prove the converse statement, $\mathfrak{D} \subseteq \mathfrak{D}(y_0)$.

Let $Z = X + iY$ be any point of $\mathfrak{D}$. Thus, for all $\theta$

$$\underline{\mu}(\theta) \leqq X \cos \theta + Y \sin \theta \leqq \bar{\mu}(\theta).$$

In particular, for $\theta = 0$,

$$\underline{\mu}(\theta) = \sup_{p \leqq y_0' \text{ on } \mathfrak{S}_0} \mu(p) \leqq X \leqq \inf_{p \geqq y_0' \text{ on } \mathfrak{S}_0} \mu(p),$$

and the functional $\mu$ can be extended from the manifold of polynomials $\mathfrak{M}_0$ to the linear manifold $\mathfrak{M}_1$, determined by $\mathfrak{M}_0$ and $y_0'$ in such a way that

$$\mu(y_0') = X.$$

We proceed to prove that $\mu$ can be now extended to the linear manifold $\mathfrak{M}_2$ determined by $\mathfrak{M}_1$ and $y_0''$ in such a way that

$$\mu(y_0'') = Y.$$

In the proof we assume that

$$\mathfrak{M}_0 \subset \mathfrak{M}_1 \subset \mathfrak{M}_2 .$$

(The treatment of exceptional cases where some of these manifolds coincide may be left to the reader.) Since every element of $\mathfrak{M}_1$ is uniquely represented as $p + \alpha y_0$, we have only to prove, in view of Introduction, **4**, that

$$m = \sup_{p + \alpha y_0' \leqq y_0'' \text{ on } \mathfrak{S}_0} \mu(p) \leqq Y \leqq \inf_{p + \alpha y_0' \geqq y_0'' \text{ on } \mathfrak{S}_0} \mu(p) = M.$$

Given any two real numbers $\xi$, $\eta$, put

$$\underline{\mu}(\xi, \eta) = \sup_{p \leqq y_0' \xi + y_0'' \eta \text{ on } \mathfrak{S}_0} \mu(p),$$

$$\bar{\mu}(\xi, \eta) = \inf_{p \geqq y_0' \xi + y_0'' \eta \text{ on } \mathfrak{S}_0} \mu(p).$$

We can characterize the domain $\mathfrak{D}$ as the common part of all closed strips

$$\underline{\mu}(\xi, \eta) \leqq X\xi + Y\eta \leqq \bar{\mu}(\xi, \eta).$$

Thus, given $\epsilon > 0$, there exists a polynomial $p_\epsilon$ and a real number $\alpha_\epsilon$ such that

$$p_\epsilon + \alpha_\epsilon y_0' \leqq y_0'' \text{ on } \mathfrak{S}_0, \qquad \mu(p_\epsilon + \alpha_\epsilon y_0') > m - \epsilon.$$

But for all $(\xi, \eta)$

$$\underline{\mu}(\xi, \eta) \leqq X\xi + Y\eta \leqq \bar{\mu}(\xi, \eta).$$

For $(\xi, \eta) = (-\alpha_\epsilon, 1)$ we have

$$\sup_{p \leqq -\alpha_\epsilon y_0' + y_0'' \text{ on } \mathfrak{S}_0} \mu(p) = \underline{\mu}(-\alpha_\epsilon, 1) \leqq -\alpha_\epsilon X + Y,$$

and since $p_\epsilon \leqq -\alpha_\epsilon y_0' + y_0''$ on $\mathfrak{S}_0$, we have $\mu(p_\epsilon) \leqq -\alpha_\epsilon X + Y$, whence

$$m - \epsilon < \mu(p_\epsilon) + \alpha_\epsilon \mu(y_0') = \mu(p_\epsilon) + \alpha_\epsilon X \leqq Y.$$

Allowing here $\epsilon \to 0$ we obtain $m \leqq Y$. By a similar argument we deduce that $Y \leqq M$. We have thus proved that the functional $\mu$ can be extended so that $\mu(y_0') = X$, $\mu(y_0'') = Y$. A repetition of the arguments used on page 3 shows the existence of a solution $\Phi$ of our $(\mathfrak{S}_0)$ moment problem such that

$$X = \mu(y_0') = \int_{\mathfrak{R}} y_0' \, d\Phi, \qquad Y = \mu(y_0'') = \int_{\mathfrak{R}} y_0'' \, d\Phi,$$

which proves that the point $X \cos \theta + Y \sin \theta$ actually belongs to $\mathfrak{D}(y_0)$.

**4.** We now concentrate our attention on the Hamburger moment problem (1.6) and on the corresponding functional (1.7) and its extension. We assume that a solution exists. Take the complex-valued function $\dfrac{1}{z - t}$ and put

$$I(z; \psi) = \int_{-\infty}^{\infty} \frac{d\psi(t)}{z - t}. \tag{1.23}$$

The previous discussion shows that a necessary condition in order that the problem (1.6) be determined is that the domain $\mathfrak{D}\left(\dfrac{1}{z - t}\right)$ reduces to a single point for each non-real $z$. We have, however, a stronger

THEOREM 1.9. *In order that the problem* (1.6) *be determined it is necessary and sufficient that the domain* $\mathfrak{D}\left(\dfrac{1}{z - t}\right)$ *consist of a single point for a sequence* $\{z_n\}$ *of values of* $z$ *which has a non-real limit point* $z_0$.

The condition of the theorem is clearly necessary. To prove that it is also sufficient assume that $\mathfrak{D}\left(\dfrac{1}{z_n - t}\right)$ reduces to a single point, for a sequence $\{z_n\}$ of values of $z$ such that $z_n \to z_0$, and $z_0$ is not real. Let $\psi_1(t)$, $\psi_2(t)$ be two solutions of (1.6). Assume for definiteness $z_0 = x_0 + iy_0$, $y_0 > 0$, and consider two functions of $z$, $I(z; \psi_1)$, $I(z; \psi_2)$, analytic in the upper half-plane $\mathfrak{I}z > 0$. Since

$z_0$ is in the interior of their domain of analyticity, and since by assumption $I(z_n, \psi_1) = I(z_n, \psi_2)$, it follows from the Vitali theorem that $I(z, \psi_1) = I(z, \psi_2)$ for $\Im z > 0$. Since, by Introduction, **5**, $\psi$ is determined by $I(z; \psi)$ substantially uniquely, it follows that the difference $\psi_1 - \psi_2$ is substantially a constant, so that the problem (1.6) is determined. As we shall see later (II, **13**), the condition of the theorem could be replaced by the condition that $\mathfrak{D}\left(\frac{1}{z_0 - t}\right)$ consist of a single point for just one non-real value $z_0$ of $z$.

Let again $z = x + iy$, $y \geqq a > 0$. We have

$$I(z; \psi) = \int_{-\infty}^{\infty} \frac{d\psi(t)}{z - t} = \int_{-\infty}^{\infty} d\psi(t)\left[\frac{1}{z} + \frac{t}{z^2} + \cdots + \frac{t^{n-1}}{z^n} + \frac{t^n}{z^n(z - t)}\right]$$

$$= \frac{\mu_0}{z} + \frac{\mu_1}{z^2} + \cdots + \frac{\mu_{n-1}}{z^n} + R_n(z; \psi),$$

where

$$R_n(z; \psi) = z^{-n} \int_{-\infty}^{\infty} \frac{t^n\, d\psi(t)}{z - t}.$$

On introducing the "absolute moments" of $\psi(t)$,

$$\overset{*}{\mu}_n(\psi) = \int_{-\infty}^{\infty} |\, t\,|^n\, d\psi(t), \qquad \overset{*}{\mu}_{2n} = \mu_{2n},$$

we see that

$$|\, R_n(z; \psi)\,| \leqq \frac{1}{|\, z\,|^n y} \int_{-\infty}^{\infty} |\, t\,|^n\, d\psi(t) = \frac{\overset{*}{\mu}_n(\psi)}{|\, z\,|^n y} \leqq \frac{\overset{*}{\mu}_n(\psi)}{|\, z\,|^n a}.$$

Let $\psi_1$, $\psi_2$ be two solutions of the problem (1.6). Then for each $n$

$$|\, I(z; \psi_1) - I(z; \psi_2)\,| = |\, R_n(z; \psi_1 - \psi_2)\,| \leqq \frac{\overset{*}{\mu}_n(\psi_1) + \overset{*}{\mu}_n(\psi_2)}{|\, z\,|^n a}. \tag{1.24}$$

On putting

$$f(z) = I(z; \psi_1) - I(z; \psi_2), \qquad |\, z\,| = r,$$

$$\overset{*}{\mu}_n(\psi_1) + \overset{*}{\mu}_n(\psi_2) = m_n,$$

we see that $f(z)$ is analytic in the half-plane $\Im z > 0$ and

$$|\, f(z)\,| \leqq \frac{m_n}{r^n}\frac{1}{a}, \quad \text{for} \quad \Im z \geqq a > 0, \qquad n = 0, 1, 2, \cdots.$$

**5.** We proceed to discuss sufficient conditions on the sequence of constants $\{m_n\}$ under which we may conclude that $f(z)$ satisfying these conditions is identically zero (Cf. Ostrowski [1]). It is clear that such conditions on $\{m_n\}$ will lead to conditions on $\{\mu_n\}$ sufficient that the problem (1.6) be determined.

Without loss of generality we may assume $a = \frac{1}{2}$, $m_0 = \frac{1}{2}$; then $f(z)$ is analytic in $\Im z > 0$ and $|\, f(z)\,| \leqq 2m_0 = 1$ for $\Im z \geqq \frac{1}{2}$. The line $y = 1/2$ is represented by the equation $|i - z| = |z|$, and the half plane $y \geqq 1/2$ by $|i - z| \leqq |z|$. Introduce a new complex variable $x' + iy' = z' = i/z$, $z = i/z'$. Then the line above is mapped onto the circle $|z' - 1| = 1$ and the half-plane $y \geqq 1/2$ onto the interior of this circle, while the half-plane $y > 0$ becomes the half-plane $x' > 0$. Call $(\gamma)$ the interior and $\gamma$ the circumference of this circle. Dropping the primes, we are led to consider the class of functions $f(z)$ satisfying the following conditions:

(1.25)
(i) $f(z)$ is analytic in $(\gamma)$ and on $\gamma$, except perhaps at $z = 0$;
(ii) $|f(z)| \leqq 1$ in $(\gamma)$;
(iii) $|f(z)| \leqq Cm_n r^n$ in $(\gamma)$ and on $\gamma$, except perhaps at $z = 0$; $n = 1, 2, 3, \cdots$; $|z| = r$.

Introduce the function

$$T(\rho) = \sup_{n \geqq 1} \frac{\rho^n}{m_n}. \tag{1.26}$$

For some values of $\rho$, $T(\rho)$ may be infinite, and for $\rho$ sufficiently large, $T(\rho) \geqq 1$.

LEMMA 1.1. *If $f(z)$ satisfies the conditions* (1.25) *and if*

$$\int_1^\infty \log T(\rho)\rho^{-2}\, d\rho = \infty,$$

*then $f(z)$ is identically* 0.

It will be sufficient to prove that if $f(z) \not\equiv 0$ satisfies (1.25), then

$$\int_1^\infty \log T(r)r^{-2}\, dr < \infty.$$

From the definition of $T(\rho)$ it follows that for each $\rho > 0$ and $\epsilon > 0$ there exists a value of $n$ such that

$$T\left(\frac{1}{\rho}\right) - \epsilon < \frac{1}{m_n \rho^n}.$$

In view of (1.25), we see that in $(\gamma)$ and on $\gamma$, except perhaps at $z = 0$,

$$|f(z)| \leqq Cm_n r^n < C/(T(1/r) - \epsilon),$$

and since $\epsilon$ was arbitrary,

$$|f(z)| \leqq C/T(1/r), \qquad T(1/r)/C \leqq \frac{1}{|f(z)|},$$

and finally,

$$\log T(1/r) \leqq \log \frac{1}{|f(z)|} + C'$$

where $C'$ is another constant. Thus the condition

$$\int_\gamma \log 1/|f(z)|\,|dz| < \infty \quad \text{implies} \quad \int_\gamma \log T(1/r)|\,dz| < \infty.$$

By a well known theorem [Pólya und Szegö, 1, vol. I, p. 119] the assumption that $f(z) \not\equiv 0$ in $(\gamma)$ implies that

$$\int_\gamma \log (1/|f(z)|) \, |dz| < \infty,$$

so that also

$$\int_\gamma \log T(1/r) \, |dz| = 2 \int_{\gamma'} \log T(1/r) \, |dz| < \infty,$$

where $\gamma'$ is the upper half of the circle $\gamma$. Now, on $\gamma'$

$$r = |z| = 2 \cos \theta, \qquad |dz| = 2 \, |d\theta|, \qquad \left| \frac{dz}{dr} \right| = \frac{1}{\sin \theta}$$

so that

$$\begin{aligned} \int_{\gamma'} \log T(1/r) \, |dz| &= \int_{\gamma'} \log T(1/r) \, dr \left| \frac{dz}{dr} \right| \\ &\geq \int_{\theta=\pi/3}^{\theta=\pi/2} \log T(1/r) \, dr \left| \frac{dz}{dr} \right| \geq \int_0^1 \log T(1/r) \, dr \\ &= \int_1^\infty \log T(\rho) \rho^{-2} \, d\rho. \end{aligned}$$

Hence

$$\int_1^\infty \log T(\rho) \rho^{-2} \, d\rho < \infty.$$

The next step is to express the condition of Lemma 1.1 in terms of the constants $m_n$. Let $m_n^{1/n} = \beta_n$, and denote by

(1.27) $$\beta_n^* = \inf_{\nu \geq n} \beta_\nu, \qquad n = 1, 2, \cdots,$$

the "Faber minorant" of the sequence $\{\beta_n\}$.

LEMMA 1.2. *The integral* $\int_1^\infty \log T(\rho) \rho^{-2} \, d\rho$ *and the series* $\sum_{n=1}^\infty 1/\beta_n^*$ *converge or diverge simultaneously.*

If $\underline{\lim}\, \beta_n < \infty$, it is clear that $\sum_{n=1}^\infty 1/\beta_n^* = \infty$. But then there exists a constant $R$ and a sequence $\{n_k\}$ such that $\beta_{n_k} = (m_{n_k})^{1/n_k} < R$, $m_{n_k} < R^{n_k}$, and

$$\frac{\rho^{n_k}}{m_{n_k}} > \left( \frac{\rho}{R} \right)^{n_k}, \quad \text{for all} \quad \rho > 0.$$

This implies $T(\rho) = \infty$ for all $\rho > R$ so that the integral also diverges. Now assume $\beta_n \to \infty$, so that (1.27) may be written as

$$\beta_n^* = \min_{\nu \geq n} \beta_\nu .$$

It is clear that

$$\beta_n^* \leqq \beta_n \text{ for all } n, \qquad \beta_n^* \uparrow \infty,$$

$$\sum_{n=1}^{\infty} 1/\beta_n^* \geqq \sum_{n=1}^{\infty} 1/\beta_n .$$

For any $\rho > 0$ let $n(\rho)$ be the number of the $\beta_n^* \leqq \rho$. Then $\beta_{n(\rho)}^* \leqq \rho$, and $n(\rho)$ is a step function assuming only integral values and such that $n(\rho) = 0$ for $0 \leqq \rho < \beta_1^*$. For any given $\rho > 0$ there exists an integer $n_1(\rho) \geqq n(\rho)$ such that $\beta_{n(\rho)}^* = \beta_{n_1(\rho)}$. Thus

$$\beta_{n(\rho/e)}^* = \beta_{n_1(\rho/e)}, \qquad n_1(\rho/e) \equiv n_1 \geqq n(\rho/e).$$

We now have

$$T(\rho) = \sup_{n \geqq 1} \frac{\rho^n}{m_n} = \sup_{n \geqq 1} \frac{\rho^n}{\beta_n^n} \geqq \frac{\rho^{n_1}}{\beta_{n_1}^{n_1}} = \left(\frac{\rho}{\beta_{n_1}}\right)^{n_1}$$

$$= \left\{\frac{\rho}{\beta_{n(\rho/e)}^*}\right\}^{n_1} \geqq \left\{\frac{\rho}{\beta_{n(\rho/e)}^*}\right\}^{n(\rho/e)} \geqq \left(\frac{\rho}{\rho/e}\right)^{n(\rho/e)} = e^{n(\rho/e)},$$

whence

$$\log T(\rho) \geqq n(\rho/e). \tag{1.28}$$

We now introduce the function

$$V(\rho) = \int_0^\rho \frac{n(t)}{t}\, dt = n(\rho) \log \rho - \int_0^\rho \log t \, dn(t)$$

$$= n(\rho) \log \rho - \sum_{\nu=1}^{n(\rho)} \log \beta_\nu^* = \log \frac{\rho^{n(\rho)}}{\beta_1^* \beta_2^* \cdots \beta_{n(\rho)}^*},$$

and consider the sequence

$$\left\{\frac{r^n}{\beta_1^* \cdots \beta_n^*}\right\} = \left\{\frac{r}{\beta_1^*} \frac{r}{\beta_2^*} \cdots \frac{r}{\beta_n^*}\right\}.$$

Since $\beta_\nu^* \leqq \rho$ for $\nu \leqq n(\rho)$, $\beta_\nu^* > \rho$ for $\nu > n(\rho)$, and $\beta_\nu^* \uparrow \infty$, the maximum element of this sequence is

$$\frac{\rho^{n(\rho)}}{\beta_1^* \beta_2^* \cdots \beta_{n(\rho)}^*}.$$

On the other hand, for each $n$,

$$\beta_1^* \beta_2^* \cdots \beta_n^* \leqq (\beta_n^*)^n \leqq \beta_n^n = m_n,$$

$$\frac{\rho^n}{\beta_1^* \beta_2^* \cdots \beta_n^*} \geqq \frac{\rho^n}{m_n}.$$

Hence

$$\sup_{n \geqq 1} \frac{\rho^n}{\beta_1^* \beta_2^* \cdots \beta_n^*} = \frac{\rho^{n(\rho)}}{\beta_1^* \beta_2^* \cdots \beta_{n(\rho)}^*} \geqq \sup_{n \geqq 1} \frac{\rho^n}{m_n} = T(\rho),$$

so that

$$V(\rho) = \log \frac{\rho^{n(\rho)}}{\beta_1^* \beta_2^* \cdots \beta_{n(\rho)}^*} \geqq \log T(\rho). \tag{1.29}$$

We also have

$$\sum_{n=1} \frac{1}{\beta_n^*} = \int_0^\infty \rho^{-1}\, dn(\rho) = \int_0^\infty n(\rho)\rho^{-2}\, d\rho.$$

Indeed, for any $R > 1$,

$$\int_0^R \rho^{-1}\, dn(\rho) = R^{-1} n(R) + \int_0^R n(\rho)\rho^{-2}\, d\rho,$$

so that, if

$$\int_0^\infty \rho^{-1}\, dn(\rho) = \lim_{R\to\infty} \int_0^R \rho^{-1}\, dn(\rho) < \infty,$$

then also

$$\int_0^\infty n(\rho)\rho^{-2}\, d\rho < \infty. \tag{1.30}$$

Conversely, if (1.30) is satisfied, then

$$\int_0^R \rho^{-1}\, dn(\rho) - \int_0^R n(\rho)\rho^{-2}\, d\rho = \frac{n(R)}{R} = n(R) \int_R^\infty \rho^{-2}\, d\rho \leqq \int_R^\infty n(\rho)\rho^{-2}\, d\rho,$$

so that $\dfrac{n(R)}{R} \to 0$ as $R \to \infty$.

Similarly

$$\int_0^\infty \rho^{-1}\, dV(\rho) = \int_0^\infty n(\rho)\rho^{-2}\, d\rho = \int_0^\infty V(\rho)\rho^{-2}\, d\rho. \tag{1.31}$$

Now, if

$$\int_1^\infty \log T(\rho)\rho^{-2}\, d\rho < \infty,$$

then, by (1.28),

$$\int_1^\infty n(\rho/e)\rho^{-2}\, d\rho < \infty,$$

which clearly implies

$$\int_0^\infty n(\rho)\rho^{-2}\, d\rho = \int_0^\infty dn(\rho)\rho^{-1} = \sum_{n=1}^\infty 1/\beta_n^* < \infty.$$

Conversely, if $\sum_1^\infty 1/\beta_n^* < \infty$ then, by (1.30) and (1.31),

$$\int_0^\infty \rho^{-1}\, dn(\rho) = \int_0^\infty n(\rho)\rho^{-2}\, d\rho = \int_0^\infty V(\rho)\rho^{-2}\, d\rho < \infty,$$

and, by (1.29),

$$\int_1^\infty \log T(\rho)\rho^{-2}\, d\rho < \infty .$$

**6.** We are now ready to state an important result due to Carleman [1, 2, 3].

THEOREM 1.10. *A sufficient condition that the Hamburger moment problem be determined is that*

$$\sum_{n=1}^{\infty} \mu_{2n}^{-1/2n} = \infty . \tag{1.32}$$

*More generally, it is sufficient that*

$$\sum_{n=1}^{\infty} \gamma_{2n}^{-1} = \infty , \tag{1.33}$$

*where*

$$\gamma_{2n} = \inf_{\nu \geqq n} (\mu_{2\nu})^{1/2\nu}.$$

Since $\mu_{2n}^{1/2n} \geqq \gamma_{2n}$, it suffices to prove only the second condition. Now, in view of (1.24) and of Lemmas 1.1, 1.2, if $\psi_1(t)$ and $\psi_2(t)$ are any two solutions of (1.6), the function $f(z) = I(z; \psi_1) - I(z; \psi_2)$ will be identically zero, that is, the problem (1.6) will be determined, whenever the series $\sum_{n=1}^{\infty} 1/\beta_n^*$ diverges, where

$$\beta_n^* = \inf_{\nu \geqq n} [\mu_\nu^*(\psi_1) + \mu_\nu^*(\psi_2)]^{1/\nu}.$$

Since $\mu_{2n}^*(\psi_1) = \mu_{2n}^*(\psi_2) = \mu_{2n}$, it is obvious that $\beta_n^* \leqq 2\gamma_{2n}$, so that the divergence of $\sum_{n=1}^{\infty} 1/\gamma_{2n}$ implies that of $\sum_{n=1}^{\infty} 1/\beta_n^*$, and hence Theorem 1.10 is proved.

The Stieltjes moment problem

$$\mu_n = \int_0^\infty t^n\, d\varphi(t), \qquad n = 0, 1, 2, \cdots ,$$

may be considered as a Hamburger moment problem

$$\mu_n' = \int_{-\infty}^\infty t^n\, d\psi(t), \qquad n = 0, 1, 2, \cdots ,$$

where

$$\psi(t) = \begin{cases} \tfrac{1}{2}\varphi(t^2) & \text{for} \quad t > 0, \\ -\tfrac{1}{2}\varphi(t^2) & \text{for} \quad t < 0, \end{cases}$$

so that

$$\mu_{2n}' = \mu_{2n}, \qquad \mu_{2n+1}' = 0, \qquad n = 0, 1, 2, \cdots .$$

This leads to

THEOREM 1.11. *A sufficient condition that the Stieltjes moment problem be determined is that*

$$\sum_{n=1}^{\infty} \mu_n^{-1/2n} = \infty .$$

Theorems 1.10 and 1.11 contain as special cases many other results previously obtained by various authors. As examples we mention the following criteria:

$$\underline{\lim}\ \frac{1}{n} \mu_{2n}^{1/2n} < \infty, \qquad \text{[M. Riesz, 3]}$$

$$\underline{\lim}\ \frac{1}{n^{1/2}} \mu_{2n}^{1/2n} < \infty, \qquad \text{[Perron, 2]}$$

$$\overline{\lim}\ \frac{1}{n} \mu_{2n}^{1/2n} < \infty, \qquad \text{[Hamburger, 1, Stridsberg, 3]},$$

for the Hamburger moment problem, and analogous criteria for the Stieltjes moment problem.

We indicate some other consequences of Theorems 1.10 and 1.11.

COROLLARY 1.2. *If the Hamburger moment problem has a solution* $\psi(t) = \int_{-\infty}^{t} \varphi(t)\, dt$, *where* $\varphi(t) \geqq 0$ *and*

$$\int_{-\infty}^{\infty} [\varphi(t)]^q e^{\delta|t|}\, dt < \infty, \tag{1.34}$$

*for some* $q \geqq 1$ *and* $\delta > 0$, *then the problem is determined. An analogous conclusion holds in the Stieltjes case, when* (1.34) *is replaced by*

$$\int_0^{\infty} [\varphi(t)]^q e^{\delta t^{1/2}}\, dt < \infty .$$

[Hardy, 1a].

We shall discuss only the Hamburger case; an analogous discussion in the Stieltjes case is left to the reader. Observe that (1.34) with $q > 1$ implies, by an easy use of Hölder's inequality, (1.34) with $q = 1$. Assuming the latter inequality, we write

$$\mu_{2n} = \int_{-\infty}^{\infty} \varphi(t) t^{2n}\, dt = \int_{-\infty}^{\infty} \varphi(t) e^{\delta|t|} e^{-\delta|t|} t^{2n}\, dt.$$

The function $e^{-\delta|t|}t^{2n}$ is positive and even, and its maximum value is

$$e^{-2n}\left(\frac{2n}{\delta}\right)^{2n} = (\delta e)^{-2n}(2n)^{2n},$$

attained at $t = \pm\frac{2n}{\delta}$. Hence,

$$\mu_{2n} \leqq \max(e^{-\delta|t|}t^{2n}) \int_{-\infty}^{\infty} \varphi(t)e^{\delta|t|}\,dt \leqq C\left(\frac{2n}{\delta e}\right)^{2n},$$

$$\mu_{2n}^{1/2n} \leqq An,$$

and the series (1.32) diverges.

**7.** Our next application is to the theory of many-dimensional moment problems.

THEOREM 1.12. *Let the k-dimensional moment problem corresponding to the moments* $\{\mu_{j_1 j_2 \cdots j_k}\}$, $j_1, j_2, \cdots, j_k = 0, 1, 2, \cdots$, *have a solution. Let*

$$\lambda_{2n} = \mu_{2n\,0\,0\cdots 0} + \mu_{0\,2n\,0\cdots 0} + \cdots + \mu_{0\,0\cdots 0\,2n}\,.$$

*A sufficient condition for the moment problem to be determined is that*

$$\sum_{n=1}^{\infty} \lambda_{2n}^{-1/2n} = \infty. \tag{1.35}$$

As we know [Cramér, 1; Haviland, 2, 3], the $k$-dimensional distribution function $\Phi(E)$ determines, and is substantially determined by, the one-dimensional distribution function $F(u; t) \equiv \Phi(H_{t,u})$ where $H_{t,u}$ is the half-space $(t, x) = t_1x_1 + \cdots + t_kx_k \leqq u$. Thus, if $F(u; t)$ is a solution of a Hamburger moment problem which is determined, $\Phi(E)$ will be the solution of a determined moment problem. Consider the moments of the distribution function $F(u; t)$, where $t$ is any vector different from 0. We have

$$\mu_{2n}(F) = \int_{-\infty}^{\infty} u^{2n}\,d_uF(u;t) = \int_{\Re} (t, x)^{2n}\,d\Phi.$$

But, by Hölder's inequality,

$$|(t, x)| = |t_1x_1 + \cdots + t_kx_k| \leqq |t|(|x_1| + \cdots |x_k|)$$
$$\leqq |t|(x_1^{2n} + \cdots + x_k^{2n})^{1/2n}k^{1-(1/2n)}$$

Hence,

$$\mu_{2n}(F) \leqq |t|^{2n}k^{2n-1}\int_{\Re}[x_1^{2n} + \cdots + x_k^{2n}]\,d\Phi = |t|^{2n}k^{2n-1}\lambda_{2n},$$

and it is immediately seen that (1.35) implies

$$\sum_{n=1}^{\infty} [\mu_{2n}(F)]^{-1/2n} = \infty$$

for any vector $t \neq 0$. This shows that if the condition (1.35) is satisfied, the function $F(u; t)$ is substantially uniquely determined for each $t \neq 0$; hence, so is $\Phi(E)$, and the moment problem is determined. (Cf. a recent paper by Hedge [1]).

**8.** We close this chapter by quoting some examples of moment problems which are indeterminate. An easy application of the Corollary 1.2 (or of the Carleman criterion directly) to the moments

$$\mu_n = \int_{-\infty}^{\infty} t^n e^{-a|t|^\alpha}\, dt, \qquad \alpha \geqq 1, \tag{1.36}$$

$$a > 0, \quad n = 0, 1, 2, \cdots,$$

$$\mu_n = \int_0^{\infty} t^n e^{-at^\alpha}\, dt, \qquad \alpha \geqq 1/2, \tag{1.37}$$

shows that the corresponding moment problems are determined. It is readily shown that these assertions are no longer true if $\alpha < 1$ in the case (1.36) and $\alpha < 1/2$ in the case (1.37). Using the formula

$$\int_0^{\infty} y^{c-1} e^{-by}\, dy = b^{-c}\Gamma(c), \qquad c > 0, \qquad b = k + il, \qquad k > 0,$$

and putting

$$c = \frac{n+1}{\lambda}, \qquad n = 0, 1, 2, \cdots; \qquad \frac{l}{k} = \tan \alpha\pi, \qquad 0 < \alpha < 1/2, \qquad y = x^\alpha,$$

we get

$$\int_0^{\infty} x^n e^{-kx^\alpha} \sin(kx^\alpha \tan \alpha\pi)\, dx = 0, \qquad n = 0, 1, 2, \cdots.$$

Similarly, for

$$c = \frac{2n+1}{\lambda}, \qquad n = 0, 1, 2, \cdots; \qquad \frac{l}{k} = \tan \frac{\alpha\pi}{2},$$

$$0 < \alpha = \frac{2s}{2s+1}, \qquad s \text{ integer} > 0, \qquad y = x^\alpha,$$

we get

$$\int_{-\infty}^{\infty} x^n e^{-kx^\alpha} \cos\left(kx^\alpha \tan \frac{\pi\alpha}{2}\right) dx = 0, \qquad n = 0, 1, 2, \cdots.$$

This clearly shows that each of the moment problems (1.36), (1.37) has infinitely many solutions.

Another example is given by the function [Stieltjes, 5]

$$t^{-\log t}[1 + A \sin(2\pi \log t)], \qquad |A| \leqq 1,$$

since

$$\int_0^{\infty} t^{-\log t} \sin(2\pi \log t) t^n\, dt$$

$$= \int_{-\infty}^{\infty} e^{-\left(u+\frac{n+1}{2}\right)^2} \sin 2\pi\left(u + \frac{n+1}{2}\right) e^{\left(u+\frac{n+1}{2}\right)n} e^{u+\frac{n+1}{2}}\, du$$

$$= \pm e^{\left(\frac{n+1}{2}\right)^2} \int_{-\infty}^{\infty} e^{-u^2} \sin 2\pi u\, du = 0, \qquad n = 0, 1, 2, \cdots.$$

CHAPTER II

# THEORY OF THE HAMBURGER MOMENT PROBLEM.

**1.** Let $\psi(t)$ be any solution of the Hamburger moment problem

$$\mu_n = \int_{-\infty}^{\infty} t^n \, d\psi(t), \qquad n = 0, 1, 2, \cdots. \tag{2.1}$$

In the preceding chapter we have shown the fundamental role which the integral

$$I(z; \psi) = \int_{-\infty}^{\infty} \frac{d\psi(t)}{z - t}$$

plays in the discussion of the problem (2.1). In the present chapter $I(z; \psi)$ will be investigated in detail and it will be shown how the general solution of the moment problem, and also various criteria for this problem to be determined, can be obtained by constructing sequences of certain rational functions which, on the one hand, are represented by integrals of the form $I(z; \psi_n)$, where $\psi_n \to \psi$ substantially, and, on the other hand, can be developed in power series in negative powers of $z$, approximating in a certain sense the series

$$\frac{\mu_0}{z} + \frac{\mu_1}{z^2} + \cdots + \frac{\mu_n}{z^{n+1}} + \cdots, \tag{2.2}$$

which is the formal expansion of $I(z; \psi)$ in powers of $z$.

**2.** First we investigate the relationship between $I(z; \psi)$ and this power series and begin by establishing some important lemmas.

The function $-I(z; \psi)$ is a special case of a function $f(z)$ which is analytic in the half-plane $\Im z = y > 0$ and is such that $\Im f(z) \geqq 0$ in this half-plane. The following lemma gives an integral representation of such functions.

Lemma 2.1. *If $f(z)$ is analytic in the half-plane $y > 0$ and if $\Im f(z) \geqq 0$ for $y > 0$ then there exists a bounded increasing* function $\alpha(t)$ such that*

$$f(z) = Az + \int_{-\infty}^{\infty} \frac{1 + tz}{t - z} \, d\alpha(t) + c, \tag{2.3}$$

*where $A$ and $c$ are real constants, and $A \geqq 0$. Furthermore,*

$$f(z)/z \to A, \quad as \quad z \to \infty \tag{2.4}$$

*in any sector*

$$0 < \epsilon \leqq \arg z \leqq \pi - \epsilon, \qquad 0 < \epsilon < \pi/2. \tag{2.5}$$

* By "increasing" we mean non-decreasing.

Consider a function $g(\zeta)$ analytic in $|\zeta| < 1$ and such that $\Im g(\zeta) \geqq 0$ in $|\zeta| < 1$. It is well known [Herglotz, 1 and F. Riesz, 1] that such a function can be represented by the integral

$$g(\zeta) = i \int_0^{2\pi} \frac{e^{i\theta} + \zeta}{e^{i\theta} - \zeta} \, d\beta(\theta) + c,$$

where $\beta(\theta)$ is an increasing bounded function. The transformation

$$\zeta = \frac{z - i}{z + i}, \qquad z = i \frac{1 + \zeta}{1 - \zeta}, \tag{2.6}$$

maps $|\zeta| < 1$ onto $\Im z > 0$, while

$$t = -\cot \frac{\theta}{2}, \qquad \frac{t - i}{t + i} = e^{i\theta},$$

maps the circumference $|\zeta| = 1$ onto the axis of reals in the $z$-plane. If $\delta$ is an arbitrarily small positive number and if we take $g(\zeta) = f(z)$, we may write

$$f(z) = g(\zeta) = i \int_0^{\delta} \frac{e^{i\theta} + \zeta}{e^{i\theta} - \zeta} \, d\beta(\theta) + i \int_{2\pi - \delta}^{2\pi} \frac{e^{i\theta} + \zeta}{e^{i\theta} - \zeta} \, d\beta(\theta)$$

$$+ \int_{\cot \frac{\delta}{2}}^{\cot \frac{\delta}{2}} \frac{1 + tz}{t - z} \, d\alpha(t) + c,$$

where $\alpha(t) \equiv \beta(\theta)$. Letting $\delta \to 0$ we get

$$f(z) = i \frac{1 + \zeta}{1 - \zeta} \{\beta(+0) - \beta(0) + \beta(2\pi) - \beta(2\pi - 0)\} + \int_{\infty}^{\infty} \frac{1 + tz}{t - z} \, d\alpha(t) + c,$$

which, by (2.6), is the desired formula (2.3).

To prove the second part of Lemma 2.1, let $z_0$ be any fixed point in the sector (2.5). Then

$$\frac{f(z) - f(z_0)}{z - z_0} = A + \int_{\infty}^{\infty} \frac{t^2 + 1}{(t - z)(t - z_0)} \, d\alpha(t). \tag{2.7}$$

Since

$$\left| \frac{t^2 + 1}{(t - z)(t - z_0)} \right| \leqq \frac{|1 + t^{-2}|}{\sin^2 \epsilon}$$

when $z$ remains in the sector (2.5), it is readily proved by splitting the integral into two parts $\int_{|t| \leqq T} + \int_{|t| > T}$, where $T$ is arbitrarily large but fixed, that the integral term in (2.7) tends to 0 as $z \to \infty$; thus it follows that $f(z)/z \to A$.

Lemma 2.2. *If $f(z)$ is analytic in the half-plane $y > 0$ and $\Im f(z) \geqq 0$, and if, in addition, there exists a finite limit*

$$c_0 = \lim_{z \to \infty} z f(z), \qquad \epsilon \leqq \arg z \leqq \pi - \epsilon, \tag{2.8}$$

*then for* $y > 0$,

$$f(z) = \int_{-\infty}^{\infty} \frac{d\gamma(t)}{t - z},$$

*where* $\gamma(t)$ *is bounded and increasing, while* $c_0$ *must be real and*

$$\int_{-\infty}^{\infty} d\gamma(t) = -c_0 .$$

Lemma 2.1 shows that in the present case $A = 0$. Hence

$$f(z) = \int_{-\infty}^{\infty} \frac{1 + tz}{t - z} \, d\alpha(t) + c.$$

We now have

$$\Re(-iyf(iy)) = \int_{-\infty}^{\infty} \frac{y^2(1 + t^2)}{t^2 + y^2} \, d\alpha(t) = \int_{-\infty}^{\infty} \frac{1 + t^2}{1 + t^2/y^2} \, d\alpha(t).$$

Since the left-hand member tends to a finite limit as $y \to \infty$, we see that there exists a constant $M$ such that for all values of $y \geqq y_0 > 0$,

$$\int_{-\infty}^{\infty} \frac{(1 + t^2)\, d\alpha(t)}{1 + t^2/y^2} < M.$$

Let $T$ be any positive number. Then

$$\int_{-T}^{T} \frac{(1 + t^2)\, d\alpha(t)}{1 + t^2/y^2} < M,$$

and, allowing $y \to \infty$,

$$\int_{-T}^{T} (1 + t^2)\, d\alpha(t) \leqq M.$$

This shows the existence of $\int_{-\infty}^{\infty} t^2 \, d\alpha(t)$, so that the function

$$\gamma(t) = \int_{-\infty}^{t} (1 + \tau^2)\, d\alpha(\tau)$$

is bounded and increasing. Moreover, the existence of $\int_{-\infty}^{\infty} t^2 \, d\alpha(t)$ implies that of $\int_{-\infty}^{\infty} t \, d\alpha(t)$, so that we may write

$$f(z) = c - \int_{-\infty}^{\infty} t \, d\alpha(t) + \int_{-\infty}^{\infty} \frac{d\gamma(t)}{t - z} = c_1 + \int_{-\infty}^{\infty} \frac{d\gamma(t)}{t - z}.$$

But

$$z \int_{-\infty}^{\infty} \frac{d\gamma(t)}{t - z} = -\int_{-\infty}^{\infty} d\gamma(t) + \int_{-\infty}^{\infty} \frac{t}{t - z} \, d\gamma(t),$$

and since in the sector (2.5) $|t/(t - z)| < (\sin \epsilon)^{-1}$, it is readily seen that

$$z \int_{-\infty}^{\infty} \frac{d\gamma(t)}{t - z} \to - \int_{-\infty}^{\infty} d\gamma(t), \qquad \text{as } z \to \infty.$$

In view of the condition (2.8), this shows that $c_1 = 0$ and $c_0 = -\int_{-\infty}^{\infty} d\gamma(t)$.

LEMMA 2.3. *If $f(z)$ is analytic in $y > 0$, and $\Im f(z) \leqq 0$, and if in any sector*

$$\epsilon \leqq \arg z \leqq \pi - \epsilon, \qquad 0 < \epsilon < \frac{\pi}{2}, \tag{2.5}$$

*$f(z)$ admits of the representation*

$$f(z) = \alpha_0 + \frac{\beta_0}{z} + R(z),$$

*where $\alpha_0$ is real and*

$$R(z) = o(1/z), \quad \text{as} \quad z \to \infty,$$

*then either $\beta_0 = 0$ and $f(z) \equiv \alpha_0$, or $\beta_0 > 0$ and, when $z$ ranges over the half-plane $y > y_0 > 0$, $w = f(z)$ remains in the interior of the circle $C$ in the lower half-plane $\Im w \leqq 0$, of diameter $\beta_0/y_0$, which is tangent to the axis of reals at $\alpha_0$. This also holds when $z = z_0 = x_0 + iy_0$ is on the line $y = y_0$, except in the case*

$$f(z) = \alpha_0 + \frac{\beta_0}{z - \alpha_1},$$

*where $\alpha_1$ is another real constant. In this case $w = f(z)$ lies on $C$. The linear function $w = \alpha_0 + \dfrac{\beta_0}{z - \alpha_1}$ maps the half-plane $y \geqq 0$ onto the half-plane $\Im w \leqq 0$ and the straight line $y = y_0$ onto the circle $C$.*

It is clear that the condition $\Im f(z) \leqq 0$ implies $\beta_0 \geqq 0$. If now $\beta_0 = 0$, then, by Lemma 2.2, $f(z) \equiv \alpha_0$. If $\beta_0 > 0$, we must have $\Im f(z) < 0$ for $y > 0$. Then the function

$$g(z) = \frac{1}{f(z) - \alpha_0} = \frac{z}{\beta_0 + zR(z)}$$

is analytic in the half-plane $y > 0$, and satisfies the conditions

$$\Im g(z) > 0 \quad \text{for} \quad y > 0, \qquad g(z)/z \to 1/\beta_0,$$

as $z \to \infty$ in the sector (2.5). Thus, by Lemma 2.1,

$$g(z) = \frac{z}{\beta_0} + g_1(z),$$

where $\Im g_1(z) \geqq 0$ for $y > 0$. It follows that

$$\Im g(z) \geqq \Im z/\beta_0,$$

and since $f(z) = \alpha_0 + 1/g(z)$, the first part of Lemma 2.3 is proved. If at a point

$z_0 = x_0 + iy_0$, $\Im g_1(z_0) = 0$, then $g_1(z)$ reduces to a real constant, say $-\dfrac{\alpha_1}{\beta_0}$ This proves the last statement of the Lemma 2.3.

The following lemma has been proved by the preceding argument.

LEMMA 2.4. *If $f(z)$ is analytic in $y > 0$ and $\Im f(z) < 0$ for $y > 0$, and if*

$$f(z) = \alpha_0 + \frac{\beta_0}{z} + R(z),$$

*where $\beta_0 > 0$ and $R(z) = o(1/z)$ as $z \to \infty$ in the sector* (2.5), *then the function*

$$f_1(z) = \frac{\beta_0}{\alpha_0 - f(z)} + z$$

*is also analytic in $y > 0$, and $\Im f_1(z) \leqq 0$.*

**3.** In general, if we have a formal power series

$$\frac{a_0}{z} + \frac{a_1}{z^2} + \cdots + \frac{a_n}{z^{n+1}} + \cdots$$

and a function $F(z)$ analytic in an infinite domain $D$ and such that

$$R_n(z) = f(z) - \left(\frac{a_0}{z} + \cdots + \frac{a_{n-1}}{z^n}\right) = o(z^{-n}), \qquad n = 1, 2, \cdots,$$

as $z \to \infty$ remaining in $D$, we say that $F(z)$ is *asymptotically represented* in $D$ by $\sum_{n=0}^{\infty} a_n z^{-n-1}$.

The following theorems show the relationship between $I(z; \psi)$ and the power series (2.2).

THEOREM 2.1. *If $\psi(t)$ is any solution of the moment problem* (2.1), *then $I(z;\psi)$ is analytic in $y > 0$, $\Im I(z;\psi) \leqq 0$, and $I(z; \psi)$ is asymptotically represented by the series $\sum_0^\infty \mu_n z^{-n-1}$ in any sector* (2.5).

*Conversely, if $f(z)$ is analytic in $y > 0$, $\Im f(z) \leqq 0$ in $y > 0$, and if $f(z)$ is asymptotically represented by the series $\sum_0^\infty \mu_n z^{-n-1}$ in any sector* (2.5), *then there exists a unique solution $\psi(t)$ of the moment problem* (2.1), *such that $f(z) = I(z; \psi)$.*

The last relation expresses a one-to-one correspondence between the elements of the above class of functions $f(z)$ and those of the class of solutions $\psi(t)$ of the moment problem (2.1). In this correspondence all functions $\psi(t)$ which are substantially equal are not considered as distinct.

Theorem 2.1 shows that the problem of finding all solutions of the moment problem (2.1) is equivalent to that of determining all functions $f(z)$ with the following properties: (i) $f(z)$ is analytic in the half-plane $y > 0$; (ii) $\Im f(z) \leqq 0$ for $y > 0$; (iii) $f(z)$ is asymptotically represented by the series $\sum_0^\infty \mu_\nu z^{-\nu-1}$ in any sector (2.5).

An analogous situation exists in the case where the moment problem (2.1) is replaced by the "reduced moment problem"

$$\mu_\nu = \int_{-\infty}^{\infty} t^\nu \, d\psi(t), \qquad \nu = 0, 1, \cdots, 2n, \tag{2.9}$$

and the above class of functions by the class of functions $f(z)$ analytic in $y > 0$, $\Im f(z) \leqq 0$, and, in addition,

$$f(z) = \frac{\mu_0}{z} + \cdots + \frac{\mu_{2n}}{z^{2n+1}} + R_{2n+1}(z), \tag{2.10}$$

where $R_{2n+1}(z) = o(z^{-2n-1})$, as $z \to \infty$ in any sector (2.5). This is shown by

THEOREM 2.2. *If $\psi_n(t)$ is any solution of the reduced moment problem* (2.9), *then $I(z, \psi_n)$ is analytic in $y > 0$, $\Im I(z, \psi_n) \leqq 0$ in $y > 0$, and $f(z) = I(z, \psi_n)$ satisfies condition* (2.10).

*Conversely, if $f(z)$ is analytic in $y > 0$, $\Im f(z) \leqq 0$, and if, in addition,* (2.10) *is satisfied for a fixed value of $n$, then there exists a unique solution $\psi_n(t)$ of the the reduced moment problem* (2.9) *such that*

$$f(z) = I(z; \psi_n).$$

Thus we again have one-to-one correspondence between the class of functions $f(z)$ satisfying the above conditions and the class of solutions $\psi_n(t)$ of the reduced moment problem (2.9).

We shall give a proof of Theorem 2.2 only, since Theorem 2.1 can be proved in a similar and even simpler way.

Let $\psi_n(t)$ be any solution of the reduced moment problem (2.9). The facts that $I(z; \psi_n)$ is analytic in $y > 0$ and that $\Im I(z; \psi_n) \leqq 0$ in $y > 0$ are clear. Furthermore,

$$\begin{aligned} I(z; \psi_n) &= \int_{-\infty}^{\infty} \frac{d\psi_n(t)}{z - t} \\ &= \int_{-\infty}^{\infty} \left[ \frac{1}{z} + \frac{t}{z^2} + \cdots + \frac{t^{2n-1}}{z^{2n}} + \frac{t^{2n}}{z^{2n+1}} + \frac{t^{2n+1}}{z^{2n+1}(z - t)} \right] d\psi_n(t) \\ &= \frac{\mu_0}{z} + \frac{\mu_1}{z^2} + \cdots + \frac{\mu_{2n-1}}{z^{2n}} + \frac{\mu_{2n}}{z^{2n+1}} + \frac{1}{z^{2n+1}} \int_{-\infty}^{\infty} \frac{t^{2n+1} \, d\psi_n(t)}{z - t}, \end{aligned}$$

so that it remains to prove that

$$\int_{-\infty}^{\infty} \frac{t^{2n+1} \, d\psi_n(t)}{z - t} \to 0,$$

as $z \to \infty$ in a sector (2.5). But in this sector

$$\left| \frac{t}{z - t} \right| \leqq \frac{1}{\sin \epsilon}, \qquad \left| \frac{z}{z - t} \right| \leqq \frac{1}{\sin \epsilon},$$

so that for any positive $T$ we may write

$$\left| \int_{-\infty}^{\infty} \frac{t^{2n+1} \, d\psi_n(t)}{z - t} \right| \leqq \frac{1}{|z| \sin \epsilon} \int_{-T}^{T} | t^{2n+1} | \, d\psi_n(t) + \frac{1}{\sin \epsilon} \left[ \int_{-\infty}^{-T} + \int_{T}^{\infty} t^{2n} \, d\psi_n(t) \right].$$

First, choose $T$ so large that the second term on the right is less than $\delta/2$, then $|z|$ so large that the first term is also less than $\delta/2$.

To prove the converse statement of Theorem 2.2, assume that $f(z)$ is analytic in $y > 0$, that $\Im f(z) \leqq 0$ in $y > 0$ and that

$$R_{2n+1}(z) = f(z) - \sum_{\nu=0}^{2n} \mu_\nu t^{-\nu-1} = o(z^{-2n-1}),$$

as $z \to \infty$ remaining in the sector (2.5). Since $zf(z) \to \mu_0$, Lemma 2.2 shows the existence of an increasing bounded function $\psi_n(t)$ such that

$$f(z) = \int_{-\infty}^{\infty} \frac{d\psi_n(t)}{z - t}, \qquad \mu_0 = \int_{-\infty}^{\infty} d\psi_n(t).$$

It remains to prove that

$$\int_{-\infty}^{\infty} t^\nu \, d\psi_n(t) = \mu_\nu, \qquad \nu = 1, 2, \cdots, 2n. \tag{2.11}$$

Assume that formulas (2.11) have been proved for $\nu = 1, 2, \cdots, 2k$, $k < n$, and prove their validity also for $\nu = 2k + 1, 2k + 2$. Write

$$I(z; \psi_n) = \sum_{\nu=0}^{2k} \mu_\nu z^{-\nu-1} + \frac{1}{z^{2k+1}} \int_{-\infty}^{\infty} \frac{t^{2k+1} \, d\psi_n(t)}{z - t}.$$

If we put here $z = iy$ and take into consideration (2.10), we have

$$\int_{-\infty}^{\infty} \frac{t^{2k+1} \, d\psi_n(t)}{iy - t} = \frac{\mu_{2k+1}}{iy} + \frac{\mu_{2k+2}}{y^2} + o(y^{-2}). \tag{2.12}$$

On taking the real parts we see that

$$\int_{-\infty}^{\infty} \frac{t^{2k+2} \, d\psi_n(t)}{t^2 + y^2} = \frac{\mu_{2k+2}}{y^2} + o(y^{-2}),$$

$$\int_{-\infty}^{\infty} \frac{t^{2k+2} \, d\psi_n(t)}{1 + t^2/y^2} = \mu_{2k+2} + o(1),$$

and an application of the same argument as used in the proof of Lemma 2.2 shows the existence of $\int_{-\infty}^{\infty} t^{2k+2} d\psi_n(t)$ and also that this integral equals $\mu_{2k+2}$. The integral $\int_{-\infty}^{\infty} | t^{2k+1} | \, d\psi_n(t)$, and hence $\int_{-\infty}^{\infty} t^{2k+1} \, d\psi_n(t)$ also exist. On multiplying (2.12) by $iy$ and allowing $y \to \infty$, we prove lastly that

$$\int_{-\infty}^{\infty} t^{2k+1} \, d\psi_n(t) = \mu_{2k+1}.$$

Finally, the uniqueness of the solution $\psi_n(t)$ follows from the Stieltjes inversion formula (Introduction, **5**).

**4.** We thus naturally come to the following problem.

PROBLEM ($N_n$). *Determine all functions $f(z)$ which are analytic in the half-plane $y > 0$, $\mathfrak{I}f(z) \leqq 0$, and such that, for a fixed value of $n \geqq 0$,*

$$f(z) = \alpha_0 + \frac{\mu_0}{z} + \frac{\mu_1}{z^2} + \cdots + \frac{\mu_{2n}}{z^{2n+1}} + R_{2n+1}(z),$$

*where $\alpha_0$ is a real constant, and*

$$R_{2n+1}(z) = o(z^{-2n-1}),$$

*as $z \to \infty$ in any sector* (2.5).

We clearly must have $\mu_0 \geqq 0$. If $\mu_0 = 0$, then, by Lemma 2.3, $f(z) \equiv \alpha_0$, so that $\mu_1 = \cdots = \mu_{2n} = 0$. If, however, $\mu_0 > 0$, then necessarily $\mathfrak{I}f(z) < 0$ for $y > 0$, so that the function $\dfrac{\mu_0}{\alpha_0 - f(z)}$ is analytic in $y > 0$ and also has a non-positive imaginary part. Now write $\beta_0$ instead of $\mu_0$ and consider the function

$$f_1(z) = \frac{\beta_0}{\alpha_0 - f(z)} + z, \qquad f(z) = \alpha_0 + \frac{\beta_0}{z - f_1(z)}.$$

By Lemma 2.4, $f_1(z)$ is analytic in $y > 0$ and $\mathfrak{I}f_1(z) \leqq 0$. Furthermore, it is readily seen that $f_1(z)$ admits of an expansion

$$(2.13) \qquad f_1(z) = \alpha_1 + \frac{\mu_0^{(1)}}{z} + \frac{\mu_1^{(1)}}{z^2} + \cdots + \frac{\mu_{2n-2}^{(1)}}{z^{2n-1}} + R_{2n-1}^{(1)}(z),$$

where $\alpha_1, \mu_0^{(1)}, \cdots, \mu_{2n-2}^{(1)}$ are real and uniquely determined by $\alpha_0, \mu_0, \cdots, \mu_{2n}$, and

$$(*) \qquad R_{2n-1}^{(1)}(z) = o(z^{-2n+1}),$$

as $z \to \infty$ in the sector (2.5). The coefficients $\alpha_1, \mu_0^{(1)}, \cdots, \mu_{2n-2}^{(1)}$ are determined in the simplest way if we introduce the expression

$$s_{2n+1} = \sum_{\nu=0}^{2n} \mu_\nu z^{-\nu-1}$$

find the expansion of

$$-\frac{\mu_0}{s_{2n+1}} = -z + \alpha_1 + \frac{\mu_0^{(1)}}{z} + \cdots + \frac{\mu_{2n-2}^{(1)}}{z^{2n-1}} + R(z),$$

where $R(z) = o(z^{-2n+1})$, and observe that

$$f_1(z) - z = \frac{\beta_0}{\alpha_0 - f(z)} = -\frac{\beta_0}{s_{2n+1}} + \frac{\beta_0 R_{2n+1}(z)}{s_{2n+1}(f(z) - \alpha_0)}.$$

Conversely, if $f_1(z)$ is any function analytic in $y > 0$, $\mathfrak{I}f_1(z) \leqq 0$, and admits of the expansion (2.13), where $\alpha_1, \mu_0^{(1)}, \cdots, \mu_{2n-2}^{(1)}$ are determined as above, and $R_{2n-1}^{(1)}(z)$ satisfies (*), it is seen immediately that

$$f(z) = \alpha_0 + \frac{\beta_0}{z - f_1(z)}$$

is a solution of our problem ($N_n$).

The same argument can be now applied to $f_1(z)$. We must have $\mu_0^{(1)} \geqq 0$. If $\mu_0^{(1)} = 0$, then, by Lemma 2.3, $f_1(z) \equiv \alpha_1$. If $\mu_0^{(1)} > 0$, then we write $\mu_0^{(1)} = \beta_1$, and introduce the new function

$$f_2(z) = \frac{\beta_1}{\alpha_1 - f_1(z)} + z$$

analytic in $y > 0$, $\Im f_2(z) \leqq 0$, which admits of an expansion

$$f_2(z) = \alpha_2 + \frac{\mu_0^{(2)}}{z} + \cdots + \frac{\mu_{2n-4}^{(2)}}{z^{2n-3}} + R_{2n-3}^{(2)}(z),$$

where $\alpha_2$, $\mu_0^{(2)}$, $\cdots$, $\mu_{2n-4}^{(2)}$ are real and uniquely determined, while

$$R_{2n-3}^{(2)}(z) = o(z^{-2n+3}),$$

as $z \to \infty$ in the sector (2.5). We can continue this argument until either we reach a constant $\beta_{k+1} = 0$, $0 \leqq k \leqq n - 1$, while $\beta_0 > 0, \cdots, \beta_k > 0$, or until we reach the value $k = n$ with all $\beta_0, \beta_1, \cdots, \beta_n > 0$. As long as we have $\beta_k > 0$ and the functions $f_1(z), \cdots f_k(z)$, $k < n$, have been constructed, we can construct the next function

$$f_{k+1}(z) = z + \frac{\beta_k}{\alpha_k - f_k(z)}, \qquad f_k(z) = \alpha_k + \frac{\beta_k}{z - f_{k+1}(z)}$$

analytic in $y > 0$, $\Im f_{k+1}(z) \leqq 0$, which admits of an expansion

$$f_{k+1}(z) = \alpha_{k+1} + \frac{\mu_0^{(k+1)}}{z} + \cdots + \frac{\mu_{2n-2k-2}^{(k+1)}}{z^{2n-2k-1}} + R_{2n-2k-1}^{(k+1)}(z),$$

and where, in the sector (2.5),

$$R_{2n-2k-1}^{(k+1)}(z) = o(z^{-2n+2k+1}), \quad \text{for} \quad k < n,$$

while, for $k = n$,

$$f_{n+1}(z) = o(z).$$

This finally leads to

THEOREM 2.3. *In order that problem $(N_n)$ have a solution it is necessary and sufficient that, in the above algorithm, either*
(i) $\beta_0 > 0$, $\beta_1 > 0$, $\cdots$, $\beta_k > 0$, *but* $\beta_{k+1} = 0$, *for a certain* $k \leqq n - 1$, *or*
(ii) *all constants* $\beta_0, \beta_1, \cdots, \beta_n > 0$.

In the case (i) the solution is uniquely determined and is given by the continued fraction

$$f(z) = \alpha_0 + \frac{\beta_0 |}{|z - \alpha_1} - \frac{\beta_1 |}{|z - \alpha_2} - \cdots - \frac{\beta_k |}{|z - \alpha_{k+1}}.$$

This is a rational function, and the corresponding series (2.2) is not only asymptotic but actually convergent for $|z|$ sufficiently large.

In the case (ii) all solutions of problem $(N_n)$ are given by continued fractions

$$f(z) = \alpha_0 + \frac{\beta_0 |}{|z - \alpha_1} - \frac{\beta_1 |}{|z - \alpha_2} - \cdots - \frac{\beta_n |}{|z + f_{n+1}(z)},$$

where $f_{n+1}(z)$ is an arbitrary function analytic in $y > 0$ and satisfying the conditions

$$\Im f_{n+1}(z) \leqq 0, \quad \text{for } y > 0,$$
$$f_{n+1}(z) = o(z), \quad \text{as } z \to \infty \text{ in any sector (2.5).}$$

**5.** We now apply the results obtained to prove

THEOREM 2.4. *For the existence of a function $f(z)$ which is analytic in $y > 0$, satisfies the condition*

$$\Im f(z) \leqq 0, \quad \textit{for } y > 0,$$

*and is asymptotically represented by the series* $\sum_0^\infty \mu_\nu z^{-\nu-1}$ *in any sector* (2.5) *it is necessary and sufficient that either* (i) *for a certain value of* $n$,

$$\beta_0 > 0, \beta_1 > 0, \cdots, \beta_n > 0, \beta_{n+1} = 0$$

(this case has been already discussed in Theorem 2.3.), *or* (ii) *all constants* $\beta_n$, $n = 0, 1, 2, \cdots$, *are* $> 0$.* *In this case all functions $f(z)$ are asymptotically represented by the series* $\sum_0^\infty \mu_\nu z^{-\nu-1}$ *not only in any sector* (2.5), *but in every half plane* $\Im z = y \geqq y_0 > 0$.

We observe that if such an $f(z)$ exists then it is a solution of problem $(N_n)$ for all values of $n$. Hence, either for some $n$ we shall have the case (i) which needs no further discussion, or for all values of $n$ we shall have $\beta_n > 0$. Thus, if we do not have (i), condition (ii) appears as a necessary condition. To prove its sufficiency we start with an arbitrary function $f(z) = f_0(z)$ and construct the infinite sequence of functions $f_1(z), f_2(z), \cdots f_k(z), \cdots$, by the above algorithm

$$f_{k+1}(z) = z + \frac{\beta_k}{\alpha_k - f_k(z)}, \qquad k = 0, 1, 2, \cdots. \tag{2.14}$$

From the argument which has been used in the proof of Theorem 2.3 it is clear that in order to establish that $f(z)$ is represented asymptotically by the series $\sum_0^\infty \mu_\nu z^{-\nu-1}$ in any given infinite domain $D$ it suffices to prove that, for each value of $n$, or even for values of $n$ belonging to some infinite sequence $\{n_k\}$,

$$f_n(z) = o(z), \quad \text{as } z \to \infty \text{ in } D.$$

Now let $f(z)$ be a function with the properties stated in Theorem 2.4. Then, for each $n$, the corresponding function $f_n(z)$ will be analytic in the half-plane $y > 0$, will satisfy the condition $\Im f_n(z) \leqq 0$ for $y > 0$ and will admit of an expansion

$$f_n(z) = \alpha_n + \frac{\beta_n}{z} + R(z),$$

* For a better understanding of this result it should be observed that conditions (i) and (ii) are respectively equivalent to the conditions $\Delta_0 > 0, \cdots, \Delta_n > 0, \Delta_{n+1} = \Delta_{n+2} = \cdots 0$, or $\Delta_n > 0$ for all $n = 0, 1, \cdots$ [see below, footnote on p. 47], so that we have obtained the same result as that of Theorem 1.2.

where $R(z) = o(1/z)$ in any sector (2.5). But then, when $z$ ranges over the half-plane $y \geqq y_0 > 0$, by Lemma 2.3, $w = f_n(z)$ will remain in the closed circle of diameter $\beta_n/y_0$ which is in the half-plane $\Im w \leqq 0$ and is tangent to the axis of reals at the point $\alpha_n$. Thus we see that the condition $f_n(z) = o(z)$ is satisfied when $z \to \infty$ not only in any sector (2.5), but in any half-plane $y \geqq y_0 > 0$, so that the series $\sum_0^\infty \mu_\nu z^{-\nu-1}$ asymptotically represents $f(z)$ in every half-plane of this type.

To complete the proof of Theorem 2.4 it remains to prove that if $\beta_n > 0$, $n = 0, 1, 2, \cdots$, then there exists at least one function with the properties stated in Theorem 2.4. For this purpose construct the sequence of functions

$$f^{(n)}(z) = \alpha_0 + \frac{\beta_0 \;|}{|z - \alpha_1} - \frac{\beta_1 \;|}{|z - \alpha_2} - \cdots - \frac{\beta_{n-1} \;|}{|z - \alpha_n}, \qquad n = 1, 2, \cdots.$$

Instead of $\alpha_n$ we might have substituted here any function $f_n^{(n)}(z)$ analytic in $y > 0$, $\Im f_n^{(n)}(z) \leqq 0$, and which satisfies the condition $f_n^{(n)}(z) = o(z)$ in any sector (2.5). We have $\Im f_n^{(n)}(z) \leqq 0$, $y > 0$, so that, by the Vitali theorem, we can extract a subsequence $\{f^{(n_k)}(z)\}$ which converges to a limit function $f(z)$ uniformly in every bounded domain of the half-plane $y > 0$. This function $f(z)$ clearly is analytic in $y > 0$ and also satisfies the condition $\Im f(z) \leqq 0$, for $y > 0$. Hence, it remains to prove that $f(z)$ is asymptotically represented by the series $\sum_0^\infty \mu_\nu z^{-\nu-1}$ in any sector (2.5). As has been observed above, this will be proved if we show that for each $k$, $f_{n_k}(z) = o(z)$, as $z \to \infty$ in (2.5), where $f_{n_k}(z)$ is obtained from $f(z)$ by means of the algorithm (2.14). To do this, construct, by the same algorithm, for each function $f^{(n_k+\nu)}(z)$, $\nu = 1, 2, \cdots$, the corresponding function $f_{n_k}^{(n_k+\nu)}(z)$. It is clear that each of these functions is analytic for $y > 0$, that $\Im f_{n_k}^{(n_k+\nu)}(z) \leqq 0$, for $y > 0$, and that each admits of an expansion

$$f_{n_k}^{(n_k+\nu)}(z) = \alpha_{n_k} + \frac{\beta_{n_k}}{z} + R_\nu(z), \qquad \nu = 1, 2, \cdots,$$

where $R_\nu(z) = o(1/z)$, as $z \to \infty$ in (2.5). Since, as $\nu \to \infty$, $f^{(n_k+\nu)}(z) \to f(z)$ in the half-plane $y > 0$, it is obvious that $f_{n_k}^{(n_k+\nu)}(z) \to f_{n_k}(z)$ in the same half-plane. On the other hand, as $z$ ranges over any half-plane $y \geqq y_0 > 0$, by Lemma 2.3, $w = f_{n_k}^{(n_k+\nu)}(z)$ ranges over the fixed circle of diameter $\beta_{n_k}/y_0$ in the lower half-plane $\Im w \leqq 0$, which is tangent to the axis of reals at $z = \alpha_{n_k}$. The same is true for the limit function $f_{n_k}(z)$, so that $f_{n_k}(z) = o(z)$, as $z \to \infty$ in any half-plane $y \geqq y_0 > 0$, and certainly in any sector (2.5). Theorem 2.4 is now completely proved.

**6.** In the subsequent discussion, unless explicitly stated to the contrary, it will be assumed that the Hamburger moment problem (2.1) has a solution with an infinite spectrum, so that all determinants $\Delta_n > 0$. Then also all $\beta_n > 0$ [cf. foot-note on p. 47] and the algorithm (2.14) can be continued indefinitely. Consider the "generalized approximant" of the associated continued fraction,

$$\frac{\beta_0 \;|}{|z - \alpha_1} - \frac{\beta_1 \;|}{|z - \alpha_2} - \cdots - \frac{\beta_n \;|}{|z - \alpha_{n+1} - \tau},$$

where $\tau$ is any real constant, including $\tau = \pm\infty$. This is a rational fraction which, for $|z|$ sufficiently large, by Theorem 2.3, admits of an expansion

(2.15) $$\frac{\mu_0}{z} + \cdots + \frac{\mu_{2n}}{z^{2n+2}} + \frac{\mu'_{2n+1}}{z^{2n+2}} + \cdots .$$

Thus we naturally come to the problem of determining all rational fractions $\frac{p(z)}{q(z)}$ whose denominators are of degree not higher than $(n+1)$, and which admit of an expansion of the type (2.15), with the same $\mu_0, \cdots, \mu_{2n}$.

We treat this problem using a symbolic notation which is a modification of that used in the previous chapter. If $P(z)$ is any polynomial, we shall denote by $P(\mu)$ the functional $\mu(P)$. The meaning of symbols such as $\mu^n P(\mu)$, $P(\mu)Q(\mu)$ is clear. We have to distinguish between the symbol $P^n(\mu)$ which means $P[\cdots P(\mu)]$ and $P(\mu)^n$ which is simply the $n$-th power of the quantity $P(\mu)$.

Now take a polynomial $q(z)$ of degree $\leqq (n+1)$, and consider the formal power series product

$$q(z)\left(\frac{\mu_0}{z} + \frac{\mu_1}{z^2} + \cdots + \frac{\mu_{2n}}{z^{2n+1}} + \cdots\right) = p(z) + \frac{q(\mu)}{z} + \frac{\mu q(\mu)}{z^2} + \cdots,$$

where $p(z)$ represents the integral part of this product. The polynomial $p(z)$ is uniquely determined and will be called the *numerator* corresponding to the *denominator* $q(z)$. If $q(z) = a_0 + a_1 z + \cdots a_{n+1}z^{n+1}$, we have

$$p(z) = z^n a_{n+1}\mu_0 + z^{n-1}(a_{n+1}\mu_1 + a_n\mu_0) +$$
$$+ z^{n-2}(a_{n+1}\mu_2 + a_n\mu_1 + a_{n-1}\mu_0) + \cdots + (a_{n+1}\mu_n + \cdots + a_1\mu_0),$$

which also can be represented by

$$p(z) = \int_{-\infty}^{\infty} \frac{q(z) - q(t)}{z - t}\, d\psi(t),$$

where $\psi(t)$ is any solution of our moment problem or even only of the reduced moment problem

$$\int_{-\infty}^{\infty} t^\nu\, d\psi = \mu_\nu, \qquad \nu = 0, 1, \cdots, n.$$

Symbolically this may be written as

$$p(z) = \frac{q(z) - q(\mu)}{z - \mu}.$$

We also have

$$I(z;\psi) - \frac{p(z)}{q(z)} = \frac{1}{q(z)} \int_{-\infty}^{\infty} \frac{q(t)\, d\psi(t)}{z - t}.$$

If we impose the conditions

(2.16) $$q(\mu) = 0,\ \mu q(\mu) = 0, \cdots, \mu^{n-1}q(\mu) = 0,$$

the polynomial $q(z)$ will be called a *quasi-orthogonal polynomial, of order* $(n + 1)$ (associated with the sequence $\{\mu_n\}$) [M. Riesz, 3]. It will be proved that $q(z)$ must be at least of degree $n$. If $q(z)$ is of degree $(n + 1)$, then we have, for $|z|$ sufficiently large,

$$(2.17)\qquad \begin{aligned} \frac{p(z)}{q(z)} &= \frac{\mu_0}{z} + \cdots + \frac{\mu_{2n}}{z^{2n+1}} + \frac{\mu'_{2n+1}}{z^{2n+2}} + \cdots, \\ \mu'_{2n+1} &= \mu_{2n+1} - \frac{\mu^n q(\mu)}{a_{n+1}}; \end{aligned}$$

but if $q(z)$ is of degree $n$, then

$$(2.18)\qquad \frac{p(z)}{q(z)} = \frac{\mu_0}{z} + \cdots + \frac{\mu_{2n-1}}{z^{2n}} + \frac{\mu'_{2n}}{z^{2n+1}} + \cdots, \qquad \mu'_{2n} = \mu_{2n} - \frac{\mu^n q(\mu)}{a_n}.$$

If $q(z)$, in addition to (2.16), satisfies also the condition

$$(2.19)\qquad \mu^n q(\mu) = 0,$$

then it will be called the *orthogonal polynomial, of degree* $(n + 1)$, associated with the sequence $\{\mu_n\}$. Equations (2.16) and (2.19) determine the orthogonal polynomial of degree $(n + 1)$ uniquely, up to a constant factor. These equations are equivalent to the statement that

$$\int_{-\infty}^{\infty} q(t)h(t)\, d\psi(t) = 0,$$

for each polynomial $h(t)$, of degree $< n$, and for each solution of the reduced moment problem (2.9).

The orthogonal polynomial of degree $n$ also satisfies conditions (2.16). Therefore, it can be considered also as a quasi-orthogonal polynomial of order $(n + 1)$, as well as one of order $n$. Since it is determined by (2.16) uniquely, up to a constant factor, it follows that the case (2.18) occurs if and only if $q(z)$ is the orthogonal polynomial of degree $n$. The sequence of orthogonal polynomials can be normalized by the condition

$$q^2(\mu) = \int_{-\infty}^{\infty} [q(t)]^2\, d\psi(t) = 1,$$

together with the requirement that the coefficient of $z^n$ be positive. The corresponding sequence of *ortho-normal polynomials* associated with the given sequence of moments $\{\mu_n\}$ is uniquely determined by the moments. In the subsequent discussions it will be denoted by $\{\omega_n(z)\}$, so that

$$(2.20)\qquad \omega_i(\mu)\omega_j(\mu) = \int_{-\infty}^{\infty} \omega_i(t)\omega_j(t)\, d\psi(t) = \delta_{ij}, \qquad i, j = 0, 1, \cdots n,$$

for any solution of the moment problem (2.1) or (2.9). The numerator $p(z)$ is a polynomial of degree one less than that of $q(z)$.

**7.** We proceed to state a few properties of quasi-orthogonal polynomials which may be immediately derived from (2.16).

LEMMA 2.5. *The degree of a polynomial satisfying conditions* (2.16) *is not less than* $n$.

If $q(z)$ is of degree $< n$, we obtain a system of $n$ homogeneous equations satisfied by the coefficients of $q(z)$, whose determinant $|\mu_{i+j}|_{i,j=0}^{n-1} > 0$, so that all coefficients vanish, and $q(z) \equiv 0$.

LEMMA 2.6. *Any three quasi-orthogonal polynomials of order* $(n + 1)$ *are linearly dependent, in other words, if* $q_1$, $q_2$ *are two linearly independent quasi-orthogonal polynomials of order* $(n + 1)$, *then any quasi-orthogonal polynomial* $q(z)$ *of order* $(n + 1)$ *can be expressed as*

$$q(z) = A_1q_1(z) + A_2q_2(z).$$

*Conversely, every expression of this type is a quasi-orthogonal polynomial of order* $(n + 1)$. *In particular, every quasi-orthogonal polynomial of order* $(n + 1)$ *can be expressed as a linear combination of two orthogonal polynomials, of degrees* $n$ *and* $(n + 1)$. *Thus, the general quasi-orthogonal polynomial depends only on two parameters.*

It is clearly possible to find three constants $A_1$, $A_2$, $A_3$, not all zero, such that the polynomial $A_1q_1(z) + A_2q_2(z) + A_3q(z)$ is of degree $\leqq (n - 1)$. Since it is also quasi-orthogonal (that is, satisfies conditions (2.16)), it reduces identically to zero. The assumption that $q_1$, $q_2$ are linearly independent implies $A_3 \neq 0$.

LEMMA 2.7. (i) *Given any value* $z_0$ *(real or complex), there always exists a quasi-orthogonal polynomial* $q(z)$ *of order* $(n + 1)$ *such that* $q(z_0) = 0$. *This polynomial is completely determined up to a constant factor.* (ii) *Two linearly independent quasi-orthogonal polynomials of order* $(n + 1)$ *have no roots in common.*

Statement (i) follows from the fact that equations (2.16), together with $q(z_0) = 0$, determine $q(z)$ up to a constant factor, since the determinant $|\mu_{i+j}|_{i,j=0}^{n-1} > 0$. Statement (ii) is an immediate consequence of (i).

LEMMA 2.8. *A quasi-orthogonal polynomial* $q(z)$ *with real coefficients has all roots real and distinct. A quasi-orthogonal polynomial* $q(z)$ *whose coefficients can not be made real by dividing by a common constant factor has all roots either in the half-plane* $y > 0$ *or in the half-plane* $y < 0$.

To prove the first part of Lemma 2.8 it is sufficient to show that $q(z)$ has at least $n$ distinct real roots of odd multiplicities. If this were not the case, it would have been possible to construct a polynomial $h(z)$, of degree $\leqq (n - 1)$, such that $q(z)h(z)$ is always $\geqq 0$. In view of the positiveness of the functional $\mu$, then also $q(\mu)h(\mu) > 0$, in contradiction with (2.16). Passing to the second part of the Lemma, we observe that a quasi-orthogonal polynomial $q(z)$ with non-real coefficients can be written as $q(z) = q_1(z) + iq_2(z)$, where $q_1$ and $q_2$ are quasi-orthogonal polynomials with real coefficients. From the first part of the lemma it follows that the ratio $\dfrac{q_1(z)}{q_2(z)}$ cannot be real at a non-real point; hence, the imaginary part of this ratio maintains a constant sign in the half-plane $y > 0$

and the opposite constant sign in the half-plane $y < 0$; this shows that $q_1(z) + iq_2(z)$ can vanish in one of these two half-planes only.

LEMMA 2.9. *The zeros of any two real linearly independent quasi-orthogonal polynomials of the same order are mutually separated.*

Let $x_0, x_1, \cdots, x_m$, $m = n$ or $n - 1$, be the roots of the real quasi-orthogonal polynomial $q_2(z)$. Then if $q_1(z)$ is another real quasi-orthogonal polynomial, we have the expansion

$$\frac{q_1(z)}{q_2(z)} = Az + B + \sum_{i=0}^{m} \frac{\gamma_i}{z - x_i}, \qquad A, B, \gamma_i \text{ real.}$$

Since, by the preceding proof, $\Im \dfrac{q_1(z)}{q_2(z)}$ is of opposite constant signs in the half-planes $y > 0$, $y < 0$, this implies that all the coefficients $\gamma_i$ must be of the same sign, which in turn implies that the roots are mutually separated, that is, between any two consecutive roots of one of these polynomials there is one and only one root of the other. Indeed, the preceding formula shows that $\operatorname{sgn} q_1(x_i + 0) = -\operatorname{sgn} q_1(x_{i+1} - 0)$.

**8.** The definition (2.16) of quasi-orthogonal polynomials reveals other properties important for the construction of *formulas of approximate quadratures.** Let $q(z)$ be a quasi-orthogonal polynomial of order $(n + 1)$ and of degree $(n + 1)$. Let $x_0, x_1, \cdots, x_n$ be its roots. Put

$$q_j(t) = \frac{q(t)}{(t - x_j)q'(x_j)}, \qquad j = 0, 1, \cdots, n.$$

For any function $f(t)$ defined in $(-\infty, \infty)$ construct the corresponding Lagrange interpolation polynomial $L_n(t;f)$ which coincides with $f(t)$ at $t = x_0, x_1, \cdots, x_n$,

$$L_n(t; f) \equiv \sum_{j=0}^{n} q_j(t) f(x_j).$$

If $\psi_n(t)$ is any solution of the reduced moment problem (2.9), we may construct the corresponding approximate quadrature formula for the function $f(t)$,

$$\begin{aligned} Q_n(f) &\equiv \int_{-\infty}^{\infty} L_n(t; f)\, d\psi_n(t) = \sum_{j=0}^{n} f(x_j) \int_{-\infty}^{\infty} q_j(t)\, d\psi_n(t) \\ &= \sum_{j=0}^{n} q_j(\mu) f(x_j) = \sum_{j=0}^{n} \rho_n(x_j) f(x_j), \end{aligned}$$

where

$$\rho_n(x_j) = q_j(\mu), \qquad j = 0, 1, \cdots, n.$$

* "Approximate" quadratures seems to us more appropriate than "mechanical" quadratures, as used by European writers.

In Chapter IV we give a detailed discussion of the convergence properties of such approximate quadratures. Here we observe only that the precise formula

$$\int_{-\infty}^{\infty} G(t)\, d\psi_n(t) = G(\mu) = Q_n(G) = \sum_{j=0}^{n} \rho_n(x_j)G(x_j) \tag{2.21}$$

holds for any polynomial $G(t)$, of degree $\leqq 2n$. Indeed, for such a polynomial we have

$$G(t) = q(t)h(t) + \sum_{j=0}^{n} q_j(t)G(x_j),$$

where $h(t)$ is a polynomial of degree $\leqq (n-1)$, so that, in view of (2.16), $q(\mu)h(\mu) = 0$, and

$$G(\mu) = \sum_{j=0}^{n} q_j(\mu)G(x_j).$$

Conversely, if $q(z)$ is a polynomial of degree $(n+1)$ such that the approximate-quadrature formula (2.21) is exact for any polynomial $G(t)$ of degree not exceeding $2n$, then $q(z)$ is a quasi-orthogonal polynomial of degree $(n+1)$. Indeed, if in (2.21) we put $G(t) = t^\nu q(t)$, $\nu = 0, 1, \cdots, n-1$, we see at once that $G(x_j) = 0$, $j = 0, 1, \cdots, n$, so that $\mu^\nu q(\mu) = 0$, $\nu = 0, 1, \cdots, n-1$. If we substitute $G(t) = (q_i(t))^2$ in (2.21), we have $q_i(x_j) = \delta_{ij}$, so that

$$\rho_n(x_i) = q_i^2(\mu) > 0, \qquad i = 0, 1, \cdots, n.$$

If we substitute $G(t) = 1, t, \cdots, t^{2n}$ in (2.21), we get

$$\sum_{j=0}^{n} \rho_n(x_j)x_j^\nu = \mu_\nu, \qquad \nu = 0, 1, \cdots, 2n. \tag{2.22}$$

It should be observed that when $q(z)$ is the orthogonal polynomial of degree $(n+1)$ the approximate-quadrature formula (2.21) holds for all polynomials $G(t)$ of degree $\leqq (2n+1)$. Hence, we may add to (2.21) the relation

$$\sum_{j=0}^{n} \rho_n(x_j)x_j^{2n+1} = \mu_{2n+1}.$$

However, (2.21) can not hold for all polynomials of degree $\geqq 2n+2$, as is easily shown by considering the polynomial $[q(t)]^2$.

Expanding $\frac{p(z)}{q(z)}$ in partial fractions we have

$$\frac{p(z)}{q(z)} = \sum_{j=0}^{n} \frac{\gamma_j}{z - x_j} = \sum_{\nu=0}^{\infty} \frac{\mu'_\nu}{z^{\nu+1}},$$

where

$$\mu'_\nu = \sum_{j=0}^{n} \gamma_j x_j^\nu.$$

On the other hand, by definition (2.18),

$$\frac{p(z)}{q(z)} = \frac{\mu_0}{z} + \cdots + \frac{\mu_{2n}}{z^{2n+1}} + \frac{\mu'_{2n+1}}{z^{2n+2}} + \cdots,$$

so that $\mu'_\nu = \mu_\nu$, $\nu = 0, 1, \cdots, 2n$. On comparing this with (2.22), we see that $\gamma_j = \rho_n(x_j)$. Hence,

$$\frac{p(z)}{q(z)} = \sum_{j=0}^{n} \frac{\rho_n(x_j)}{z - x_j}. \tag{2.23}$$

Observe that from (2.23) we have

$$\rho_n(x_j) = \frac{p(x_j)}{q'(x_j)}, \qquad j = 0, 1, \cdots, n. \tag{2.24}$$

We have already proved that all $\rho_n(x_j) > 0$. This, in connection with (2.24), immediately yields

LEMMA 2.10. *If $q(z)$ is a real quasi-orthogonal polynomial of degree $(n + 1)$ and $p(z)$ the corresponding numerator, then $p(z)$ has all roots real and distinct, and the roots of $p(z)$ and $q(z)$ are mutually separated.*

**9.** The quadrature coefficients $\rho_n(x_j)$ have an extremal property which is very important for the subsequent discussion.

LEMMA 2.11. *Let $x_0$ be any real number which is not a root of the orthogonal polynomial of degree $n$; let $P_n(z)$ be any polynomial with real coefficients, of degree not exceeding $n$, such that*

$$P_n(x_0) = 1. \tag{2.25}$$

*Then*

$$\rho_n(x_0) = \min_{P_n} P_n^2(\mu) = \min_{P_n} \int_{-\infty}^{\infty} [P_n(t)]^2 \, d\psi_n(t), \tag{2.26}$$

*where $\psi_n(t)$ is any solution of the reduced moment problem (2.9)*. If $x_0$ is a root of the orthogonal polynomial of degree $n$, then*

$$\rho_{n-1}(x_0) = \min_{P_{n-1}} P_{n-1}^2(\mu) = \min_{P_{n-1}} \int_{-\infty}^{\infty} [P_{n-1}(t)]^2 \, d\psi_{n-1}(t). \tag{2.27}$$

In fact, let $q_0(z)$ be the quasi-orthogonal polynomial, of degree $(n + 1)$, whose roots are $x_0, x_1, \cdots, x_n$. Then, in view of (2.21) and (2.25),

$$P_n^2(\mu) = \sum_{j=0}^{n} \rho_n(x_j) P_n(x_j)^2 \geqq \rho_n(x_0).$$

On the other hand, on putting here $P_n(z) = \dfrac{q_0(z)}{(z - x_0)q_0'(x_0)}$, we get

$$P_n^2(\mu) = \sum_{j=0}^{n} \rho_n(x_j)\delta_{0j} = \rho_n(x_0).$$

This proof assumes that $x_0$ is not a root of the orthogonal polynomial of degree $n$. It suffices to replace $n$ by $(n - 1)$ to complete the proof in the case when $x_0$ is a root of such a polynomial.

* The value of the integral does not depend on the choice of the solution.

Lemma 2.11 leads to another important expression for $\rho_n(x_0)$ and to an extension of the definition of $\rho_n(x_0)$ for complex values of $x_0$. Introduce the sequence of ortho-normal polynomials $\{\omega_n(z)\}$. Assuming again that $z_0$ is any number (real or complex) distinct from the roots of $\omega_n(z)$ and using the notation $P_n(z)$ for any polynomial, real or complex, of degree $n$, satisfying the condition

$$P_n(z_0) = 1,$$

let us find

$$\min_{P_n} \bar{P}_n(\mu)P_n(\mu) = \min_{P_n} \int_{-\infty}^{\infty} | P_n(t) |^2 \, d\psi_n(t).$$

We have the expansions

$$P_n(z) = \sum_{\nu=0}^{n} c_\nu \omega_\nu(z), \qquad c_\nu = \omega_\nu(\mu)P_n(\mu) = \int_{-\infty}^{\infty} \omega_\nu(t)P_n(t) \, d\psi_n(t),$$

$$\overline{P_n}(\mu)P_n(\mu) = \sum_{\nu=0}^{n} | c_\nu |^2.$$

Thus, our problem is to find $\min \sum_{\nu=0}^{n} | c_\nu |^2$, for all $c_\nu$ which satisfy the relation

$$\sum_{\nu=0}^{n} c_\nu \omega_\nu(z_0) = 1. \tag{2.28}$$

Using Cauchy's inequality we get from (2.28)

$$1 \leqq \sum_{\nu=0}^{n} | c_\nu |^2 \sum_{\nu=0}^{n} | \omega_\nu(z_0) |^2,$$

where the equality sign occurs if and only if

$$c_\nu = \lambda \overline{\omega_\nu(z_0)}, \qquad \lambda = 1 \Big/ \sum_{\nu=0}^{n} | \omega_\nu(z_0) |^2.$$

This shows that

$$\overline{P_n}(\mu)P_n(\mu) = \sum_{\nu=0}^{n} | c_\nu |^2 \geqq \lambda,$$

so that $\min_{P_n} \overline{P_n}(\mu)P_n(\mu)$, under condition (2.28), equals $\lambda$, and is attained if and only if

$$P_n(z) = \sum_{\nu=0}^{n} \overline{\omega_\nu(z_0)}\omega_\nu(z) \Big/ \sum_{\nu=0}^{n} | \omega_\nu(z_0) |^2.$$

We now introduce the notation

$$K_n(z, z_0) \equiv \sum_{\nu=0}^{n} \overline{\omega_\nu(z_0)}\omega_\nu(z), \tag{2.29}$$

$$K_n(z_0) \equiv \sum_{\nu=0}^{n} | \omega_\nu(z_0) |^2 \equiv K_n(z_0, z_0), \tag{2.30}$$

and consider a modified extremum problem namely: instead of keeping fixed the value of $P_n(z_0)$ and looking for min $\overline{P_n(\mu)}P_n(\mu)$ let us fix the value of $\overline{P_n(\mu)}P_n(\mu)$ and look for the maximum of $| P_n(z_0) |$. It is clear that this maximum is the reciprocal of the minimum of the preceding problem, and therefore is equal to $1/\lambda = K_n(z_0)$. The changes in the argument which are necessary in the case when $z_0$ is a root of $\omega_n(z)$ are obvious. We thus arrive at the following important theorem.

THEOREM 2.5. *Let $\{\omega_n(z)\}$ be the sequence of ortho-normal polynomials associated with the given sequence of moments $\{\mu_n\}$. Let $z_0$ be any number, real or complex, which is not a root of $\omega_n(z)$ and let $P_n(z)$ be any polynomial, of degree $\leqq n$. Then*

$$| P_n(z_0) |^2 \leqq \frac{\overline{P_n(\mu)}P_n(\mu)}{\rho_n(z_0)},$$

*where*

$$\frac{1}{\rho_n(z_0)} \equiv K_n(z_0) = \sum_{\nu=0}^{n} | \omega_\nu(z_0) |^2, \tag{2.31}$$

*and where the equality occurs if and only if $P_n(z)$ coincides, up to a constant factor, with the polynomial $K_n(z, z_0)$.*

When $z_0$ is real, $\rho_n(z_0)$ coincides with the quadrature coefficient of the approximate-quadrature formula determined by the quasi-orthogonal polynomial of degree $(n + 1)$ which has $z_0$ for its root. Finally, if $z_0$ is real and a root of $\omega_n(z)$, the statement of the theorem holds true if $P_n$ is replaced by $P_{n-1}$. Then

$$\frac{1}{\rho_{n-1}(z_0)} = K_n(z_0) = K_{n-1}(z_0),$$

and $\rho_{n-1}(z_0)$ appears as the quadrature coefficient of the approximate-quadrature formula determined by $\omega_n(z)$.

COROLLARY 2.1. If $G_{2n}(z)$ *is a non-negative polynomial, of degree* $\leqq 2n$, *and $z_0$ is any point which is not a root of $\omega_n(z)$, then*

$$| G_{2n}(z_0) | \leqq \frac{G_{2n}(\mu)}{\rho_n(z_0)}. \tag{2.32}$$

*In case $z_0$ is a root of $\omega_n(z)$, the inequality* (2.31) *holds if $n$ is replaced by $n - 1$.*

This follows immediately from Theorem 2.5 if we observe that the general non-negative polynomial $G_{2n}(z)$ of degree $\leqq 2n$ can be represented as $G_{2n}(z) = (P_{1n}(z))^2 + (P_{2n}(z))^2 = | P_n(z) |^2$, $P_n(z) = P_{1n}(z) + iP_{2n}(z)$, where $P_{1n}(z)$, $P_{2n}(z)$ are real polynomials of degree $\leqq n$.

Observing with Stieltjes, [5], that $\psi_n(x_0 + 0) - \psi_n(x_0 - 0)$ is one of the elements of the integrand in (2.26), it is readily seen that $\rho_n(x_0) \geqq \psi_n(x_0 + 0) - \psi_n(x_0 - 0)$, the equality being attained if and only if $\psi_n(t)$ is a solution with the mass $\rho_n(x_0)$ concentrated at $x_0$, $\omega_n(x_0) \neq 0$. Now let $\psi(t)$ be any solution of the moment problem (2.1) and let $d\psi'(t) = h(t)\, d\psi(t)$, where $h(t)$ is non-negative and integrable with respect to $\psi(t)$. Applying the same reasoning as above and allowing $n \to \infty$, we readily get $\rho'(x_0) \geqq h(x_0)\rho(x_0)$, where $\rho'(x)$ corresponds to $d\psi'(t)$.

Further important properties of $\rho_n(z_0)$ follow immediately from (2.31).

THEOREM 2.6. *For a fixed z, $\rho_n(z)$ does not increase when n increases (and certainly decreases if z is not a root of $\omega_n(z)$). Hence, $\rho(z) = \lim_{n\to\infty} \rho_n(z) \geqq 0$ exists for all z. For a fixed n, $\rho_n(z)$ decreases when $y = \Im z$ increases in absolute value, while $x = \Re z$ is kept fixed.*

The first part of Theorem 2.6 is obvious. The second part also can be readily established if we remember (Lemma 2.8) that

$$\omega_\nu(z) = a_\nu \prod_{j=1}^{\nu} (z - x_j), \qquad a_\nu > 0,\ x_j \text{ real},$$

so that

$$| \omega_\nu(x + iy) |^2 = a_\nu \prod_{j=1}^{\nu} [(x - x_j)^2 + y^2]$$

is an increasing function of $| y |$. The same holds for $K_n(z) = \sum_{\nu=0}^{n} | \omega_\nu(z) |^2 = 1/\rho_n(z)$.

**10.** Formula (2.22) shows that the monotonic function $\psi_n(t)$ whose spectrum reduces to the finite set of points $x_0, x_1, \cdots, x_n$, with jumps $\rho_n(x_j)$ at $t = x_j$, is a solution of the reduced moment problem (2.9). This leads to the following

DEFINITION. *A solution $\psi_n(t)$ of the reduced moment problem* (2.9), *whose spectrum consists of precisely $(n + 1)$ points $x_0, x_1, \cdots, x_n$, is called a distribution function of order n, associated with the given sequence of moments $\{\mu_n\}$. A distribution function of order $n + k$, $k > 0$, or a solution of the moment problem* (2.1) *will be called simply a distribution function of order higher than n.**

LEMMA 2.12. *A distribution function of order n is uniquely determined by a single point of its spectrum. To each real value $x_0$ there corresponds one and only one distribution function of order n or of order $(n - 1)$, which contains $x_0$ in its spectrum; it is of order n, if $x_0$ is not a root of the orthogonal polynomial of degree n, and of order $(n - 1)$ otherwise. There is one to one correspondence between the distribution functions of order n and quasi-orthogonal polynomials of degree $(n + 1)$. The jumps (or the concentrated masses) of any distribution function of order n coincide with the quadrature coefficients $\rho_n(x_j)$ of the approximate quadrature formula determined by the corresponding quasi-orthogonal polynomial of degree $(n + 1)$.*

The proof is obvious.

* According to Hamburger's definition, $\psi_n(t)$ should be called a distribution function of order $(n + 1)$.

**11.** We now turn to important inequalities stated by Tchebycheff [1] and proved almost simultaneously and by the same method by A. Markoff [1] and Stieltjes [3].

LEMMA 2.13. (TCHEBYCHEFF INEQUALITIES). *Let $x_j$ be a point of the spectrum $x_0 < x_1 < \cdots < x_n$ of any distribution function $\psi_n(t)$ of order $n$; let $\psi(t)$ be a distribution function of order $\geqq n$, distinct from $\psi_n(t)$. Let these two functions be normalized by the condition $\psi_n(-\infty) = \psi(-\infty) = 0$. Then*

$$\psi_n(x_j - 0) < \psi(x_j - 0) \leqq \psi(x_j + 0) < \psi_n(x_j + 0). \tag{2.33}$$

*Here we have to write*

$$\psi_n(x_j - 0) \leqq \psi(x_j - 0) \quad \text{or} \quad \psi(x_j + 0) \leqq \psi_n(x_j + 0), \tag{2.34}$$

*according as $x_j = x_0$ or $x_j = x_n$.*

The **inequalities** (2.34) are obvious because we always have

$$0 \leqq \psi(t) \leqq \mu_0, \qquad -\infty \leqq t \leqq \infty,$$

$$\psi_n(t) = \begin{cases} 0 & \text{for } t < x_0, \\ \mu_0 & \text{for } t > x_n. \end{cases}$$

Hence, in the following discussion we may assume $x_j \neq x_0, x_n$. Construct a polynomial $P_{2n}(t)$, of degree $\leqq 2n$, which satisfies the following conditions

$$P_{2n}(x_0) = P_{2n}(x_1) = \cdots = P_{2n}(x_j) = 1, \qquad P_{2n}(x_{j+1}) = \cdots = P_{2n}(x_n) = 0,$$

$$P'_{2n}(x_0) = \cdots = P'_{2n}(x_{j-1}) = P'_{2n}(x_{j+1}) = \cdots = P'_{2n}(x_n) = 0.$$

It is readily seen that

$$\text{in } (-\infty, x_j], \quad P_{2n}(t) \geqq 1, \text{ and } P_{2n}(t) > 1, \quad \text{if } t \neq x_0, x_1, \cdots, x_j;$$

$$\text{in } [x_j, \infty), \quad P_{2n}(t) \geqq 0, \text{ and } P_{2n}(t) > 0, \quad \text{if } t \neq x_{j+1}, \cdots, x_n.$$

By the quadrature formula (2.21),

$$\int_{-\infty}^{\infty} P_{2n}(t)\, d\psi_n(t) = \rho_n(x_0) + \cdots + \rho_n(x_j) = \psi_n(x_j + 0).$$

On the other hand, since the order of $\psi(t)$ is not less than that of $\psi_n(t)$,

$$\int_{-\infty}^{\infty} P_{2n}(t)\, d\psi_n(t) = \int_{-\infty}^{\infty} P_{2n}(t)\, d\psi(t).$$

Let $A \geqq 0$ be the jump of $\psi(t)$ at $t = x_j$. Using the fact that the spectra of $\psi_n(t)$ and $\psi(t)$ are distinct, it is readily seen that

$$\psi_n(x_j + 0) = \int_{-\infty}^{\infty} P_{2n}(t)\, d\psi(t) = \int_{-\infty}^{x_j - 0} P_{2n}(t)\, d\psi(t) + A + \int_{x_j + 0}^{\infty} P_{2n}(t)\, d\psi(t)$$

$$> \int_{-\infty}^{x_j - 0} d\psi(t) + A = \psi(x_j + 0),$$

which gives the desired inequality $\psi(x_j + 0) < \psi_n(x_j + 0)$. The other inequality, $\psi_n(x_j - 0) < \psi(x_j - 0)$, may be established in a similar way, by constructing a new polynomial $P_{2n}(t)$ satisfying the conditions

$$P_{2n}(x_0) = \cdots = P_{2n}(x_{j-1}) = 0, P_{2n}(x_j) = \cdots = P_{2n}(x_n) = 1,$$
$$P'_{2n}(x_0) = \cdots = P'_{2n}(x_{j-1}) = P'_{2n}(x_{j+1}) = \cdots = P'_{2n}(x_n) = 0.$$

The following important corollaries are deduced immediately from Lemma 2.13.

COROLLARY 2.2. *If $(a, b)$ is an interval of constancy of the function $\psi(t)$, then $\psi_n(t)$ can have no more than one point of its spectrum in the closed interval $[a, b]$.*

COROLLARY 2.3. *Let $x_0$ be a fixed real number, and denote by*

$$\cdots < x_{n,-r} < x_{n,-r+1} < \cdots < x_{n,0} = x_0 < x_{n,1} < \cdots < x_{n,s} < \cdots$$

*the spectrum of the distribution function of order $n$ which has a jump at $x_0$. If, for a certain value of $n$, the points $x_{n,-r}$, $x_{n,s}$ exist, they will exist for all values $n + 1$, $n + 2$, $\cdots$, and*

$$x_{n,-r} < x_{n+1,-r} < \cdots < x_0 ,$$
$$x_{n,s} > x_{n+1,s} > \cdots > x_0 ,$$

*so that*

$$\lim_{n\to\infty} x_{n,-r} \leqq x_0 , \qquad \lim_{n\to\infty} x_{n,s} \geqq x_0 .$$

COROLLARY 2.4. *The quantity $\rho_n(x_0)$ is the largest mass which can be concentrated at a given point $x_0$ by any solution of a reduced moment problem of order $\geqq n$ (or $\geqq n - 1$, if $x_0$ is a root of $\omega_n(z)$).*

*If $\psi'_n(t)$, $\psi''_n(t)$ are any two solutions of the reduced moment problem* (2.9), *then*

$$| \psi'_n(t) - \psi''_n(t) | \leqq \rho_n(t).$$

*If $t$ lies in $(c, d)$, where $c$ and $d$ are any two successive roots of $\omega_n(z)$, then*

$$| \psi'_n(t) - \psi''_n(t) | \leqq \rho_n(c) + \rho_n(d).$$

*If $\psi'(t)$, $\psi''(t)$ are any two solutions of the moment problem* (2.1), *then*

$$| \psi'(x_0) - \psi''(x_0) | \leqq \rho(x_0) = \lim_{n\to\infty} \rho_n(x_0). \tag{2.35}$$

*The largest mass which can be concentrated at a given point $x_0$ by any solution of the moment problem* (2.1) *is $\rho(x_0)$, and there always exists a solution which has the mass $\rho(x_0)$ concentrated at $x_0$. This solution is uniquely determined so that the mass concentrated at $x_0$ by any other solution will be actually less than $\rho(x_0)$.*

The statements of the Corollary 2.4, except for the two last ones, follow immediately from

$$\psi(x_0 + 0) - \psi(x_0 - 0) < \psi_n(x_0 + 0) - \psi_n(x_0 - 0) = \rho_n(x_0)$$

and from the fact (Theorem 2.6) that $\lim_{n\to\infty} \rho_n(x_0) = \rho(x_0)$ exists. To prove the existence of a solution with the mass $\rho(x_0)$ concentrated at $x_0$ it suffices to apply the First Helly Theorem (Introduction, **3**) to the sequence of functions $\{\psi_n(t)\}$. We may extract a subsequence $\{\psi_{n_k}(t)\}$ which converges substantially to a monotonic function $\psi(t)$, a solution of the moment problem (2.1), so that

$$\psi(x_0 + 0) - \psi(x_0 - 0) \leqq \rho(x_0).$$

On the other hand, let $\delta > 0$ be such that $x_0 \pm \delta$ are points of continuity of $\psi(t)$. Then

$$\begin{aligned}\psi(x_0 + \delta) - \psi(x_0 - \delta) &= \lim_k [\psi_{n_k}(x_0 + \delta) - \psi_{n_k}(x_0 - \delta)] \\ &\geqq \lim_k \rho_{n_k}(x_0) = \rho(x_0),\end{aligned}$$

and, letting $\delta \to 0$,

$$\psi(x_0 + 0) - \psi(x_0 - 0) \geqq \rho(x_0).$$

It remains to prove that the solution $\psi(t)$ for which $\psi(x_0 + 0) - \psi(x_0 - 0) = \rho(x_0)$ is uniquely determined. Assume that there exist two distinct solutions, $\psi(t)$ and $\psi'(t)$, satisfying this condition. Let $\psi_1(t)$ and $\psi_1'(t)$ be the functions obtained from $\psi(t)$ and $\psi'(t)$, respectively, by removing the mass $\rho(x_0)$ at the point $x_0$. The resulting moments will be obviously

$$\mu_n^{(1)} = \mu_n - x_0^n \rho(x_0), \qquad n = 0, 1, 2, \cdots .$$

Let $\rho^{(1)}(x_0)$ be the maximum mass which can be concentrated at $x_0$ by any solution of this new moment problem. We shall prove that $\rho^{(1)}(x_0) = 0$. In view of the criterion to be proved later (Theorem 2.9), this will imply that the new moment problem is determined, that is, the solutions $\psi_1(t)$ and $\psi_1'(t)$ are substantially equal, which occurs if and only if the original solutions $\psi(t)$ and $\psi'(t)$ are substantially the same, while we have assumed that they were distinct. Now, if we assume that $\rho^{(1)}(x_0) > 0$, by the part of Corollary 2.4 which has already been proved, there exists a solution $\psi_1''(t)$ of the new moment problem with the mass $\rho^{(1)}(x_0) > 0$ concentrated at $x_0$. If we place back the mass $\rho(x_0)$ at this point we obtain a solution $\psi''(t)$ of the original moment problem which has a mass $\rho^{(1)}(x_0) + \rho(x_0) > \rho(x_0)$ concentrated at $x_0$, which, as we know is not possible. Corollary 2.4 now follows directly.

COROLLARY 2.5. *The sequence $\{\psi_n(t)\}$ of distribution functions of order $n$, which all have a mass at a fixed point $x_0$, converges, as $n \to \infty$, to the solution $\psi(t)$ of problem* (2.1) *which has the maximum possible mass $\rho(x_0)$ at $x_0$. In particular, this holds for the sequence of distribution functions corresponding to the sequence of quasi-orthogonal polynomials* $\left\{\omega_n(x) - \dfrac{\omega_n(\lambda)}{\omega_{n-1}(\lambda)}\, \omega_{n-1}(x)\right\}$, *where $\lambda$ is fixed and $\omega_n(\lambda) \neq 0$ for $n \geqq n_0$.*

COROLLARY 2.6. *If $\psi(x)$ is the solution of the determined moment problem* (2.1), *then $\rho(x) = 0$ at all points of continuity of $\psi(x)$ and equals the jump of $\psi(x)$ at a* *point of discontinuity. In particular, if* $\psi(x)$ *is the solution of the moment problem for a finite interval* $(a, b)$, *then* $\rho(x) = 0$ *outside* $(a, b)$.

**12.** We now return to the function

$$f(z) = \int_{-\infty}^{\infty} \frac{d\psi(t)}{z - t} = I(z; \psi),$$

where $\psi(t)$ is any solution of the moment problem (2.1). We still maintain the assumption $\Delta_n > 0$, $n = 0, 1, 2, \cdots$. We have seen that $f(z)$ is asymptotically represented by the series $\sum_0^\infty \mu_\nu z^{-\nu-1}$ in any half-plane $\mathfrak{I}z = y \geqq y_0 > 0$, and that, for each value of $n$, $f(z)$ is a solution of the problem $(N_n)$ and hence is representable in the form

$$f(z) = \frac{\beta_0}{|z - \alpha_1} - \frac{\beta_1}{|z - \alpha_2} - \cdots - \frac{\beta_n}{|z - f_{n+1}(z)},$$

where $f_{n+1}(z)$ is analytic in the half-plane $y > 0$, $\mathfrak{I}f_{n+1}(z) \leqq 0$, and $f_{n+1}(z) = o(z)$, as $z \to \infty$ in any half-plane $y \geqq y_0 > 0$. Now consider the associated continued fraction

$$\frac{\beta_0}{|z - \alpha_1} - \frac{\beta_1}{|z - \alpha_2} - \cdots - \frac{\beta_n}{|z - \alpha_{n+1}} - \cdots, \tag{2.36}$$

and let

$$\mathcal{C}_n = \frac{P_n(z)}{Q_n(z)},$$

where $Q_n(z)$ is a polynomial of degree $n$ with the highest coefficient 1, be its $n$-th approximant. From the general theory of continued fractions, [Perron, 1] we know that

$$\frac{P_{n+1}(z)}{Q_{n+1}(z)} - \frac{P_n(z)}{Q_n(z)} = \frac{\beta_0 \cdots \beta_n}{Q_n(z) Q_{n+1}(z)}.$$

The right-hand member can be expanded, for sufficiently large $|z|$, in a series of negative powers of $z$, starting with the term $\dfrac{\beta_0 \cdots \beta_n}{z^{2n+1}}$. This shows that the expansion of $\dfrac{P_n(z)}{Q_n(z)}$ coincides in terms $z^{-1}, \cdots, z^{-2n}$ with the formal series $\sum_0^\infty \mu_\nu z^{-\nu-1}$, which, in turn, implies that $Q_n(z)$ is the orthogonal polynomial of degree $n$ and differs only by a constant factor from $\omega_n(z)$. Since the polynomial

$$Q(z) = \frac{1}{\Delta_n} \left| \mu_i \mu_{i+1} \cdots \mu_{i+n} z^i \right|_0^{n+1}$$

satisfies the conditions $\mu^\nu Q(\mu) = 0$, $\nu = 0, 1, \cdots, n$, and has the highest coefficient 1, we have

$$Q_{n+1}(z) = \frac{1}{\Delta_n} \left| \mu_i \mu_{i+1} \cdots \mu_{i+n} z^i \right|_0^{n+1}.$$

If $a_n$ is the highest coefficient of the ortho-normal polynomial $\omega_n(z)$, it now follows that $a_n = (\Delta_{n-1}/\Delta_n)^{\frac{1}{2}}$. It is also readily found [Perron, 1] that*

$$\beta_n = \frac{a_{n-1}^2}{a_n^2} = \frac{\Delta_{n-2}\Delta_n}{\Delta_{n-1}^2}, \qquad n \geqq 2; \qquad \beta_1 = \frac{a_0^2}{a_1^2} = \frac{\Delta_1}{\Delta_0^2}, \qquad \beta_0 = \frac{1}{a_0^2} = \Delta_0,$$

$$\Delta_n = \beta_0\beta_1 \cdots \beta_n .$$

The constants $\alpha_n$ in (2.36) are also easy to determine [Perron, 1]

The polynomials $P_n(z)$, $Q_n(z)$ satisfy the recurrence relations

$$\begin{aligned} Q_{n+1}(z) &= (z - \alpha_{n+1})Q_n(z) - \beta_n Q_{n-1}(z), \\ & \qquad\qquad\qquad\qquad\qquad n = 0, 1, 2, \cdots \\ P_{n+1}(z) &= (z - \alpha_{n+1})P_n(z) - \beta_n P_{n-1}(z), \end{aligned}$$

where

$$Q_0(z) = 1,\ Q_{-1}(z) = 0,\ P_0(z) = 0,\ P_{-1}(z) = -1.$$

It is also known from the theory of continued fractions that

$$P_{n+1}(z)Q_n(z) - Q_{n+1}(z)P_n(z) = \beta_0 \cdots \beta_n . \tag{2.37}$$

With this notation, the general solution of problem $(N_n)$ can be written in the form

$$f(z) = \frac{[z - f_{n+1}(z)]P_n(z) - \beta_n P_{n-1}(z)}{[z - f_{n+1}(z)]Q_n(z) - \beta_n Q_{n-1}(z)},$$

or, replacing $f_{n+1}(z)$ by $\alpha_{n+1} + (f_{n+1} - \alpha_{n+1})$, in the form

$$f(z) = \frac{P_{n+1}(z) - [f_{n+1}(z) - \alpha_{n+1}]P_n(z)}{Q_{n+1}(z) - [f_{n+1}(z) - \alpha_{n+1}]Q_n(z)}. \tag{2.38}$$

In particular, the generalized approximants, that is, the continued fractions

$$\frac{\beta_0 |}{|z - \alpha_1} - \frac{\beta_1 |}{|z - \alpha_2} - \cdots - \frac{\beta_n |}{|z - \alpha_{n+1} - \tau}, \qquad \tau \text{ a real parameter},$$

are represented by

$$\frac{P_{n+1}(z) - \tau P_n(z)}{Q_{n+1}(z) - \tau Q_n(z)} \equiv \mathcal{C}_{n+1}(z; \tau). \tag{2.39}$$

On the other hand, the general quasi-orthogonal polynomial of degree $(n + 1)$, with real coefficients, can be written as $Q_{n+1}(z) - \tau Q_n(z)$, and the corresponding numerator as $P_{n+1}(z) - \tau P_n(z)$; thus, it is clear that to each real value of $\tau$ in

* These formulas show that the condition $\beta_n > 0$, $n = 1, 2, \cdots$ implies $\Delta_n > 0$, $n = 0, 1, 2, \cdots$, and conversely. Finally, the conditions $\beta_0 > 0, \cdots, \beta_n > 0$, $\beta_{n+1} = \beta_{n+2} = \cdots = 0$ imply the existence of the solution of the moment problem which is unique and has precisely $(n + 1)$ points in its spectrum. Then we know that $\Delta_0 > 0, \cdots, \Delta_n > 0$, $\Delta_{n+1} = \Delta_{n+2} = \cdots = 0$. The converse is also readily verified. This gives a direct proof of the equivalence of the conditions of Theorems 1.2 and 2.3.

(2.39), admitting also the value $\tau = \infty$, there corresponds just one fraction $\frac{p(z)}{q(z)}$, where $q(z)$ ranges over the class of all quasi-orthogonal polynomials of order $(n + 1)$, and conversely.

**13.** Let $z$ be a fixed point in the half-plane $y > 0$. Consider two complex variables $\zeta$ and $\zeta_{n+1}$ related by the linear transformation formula

$$\zeta = \frac{P_{n+1}(z) - \zeta_{n+1} P_n(z)}{Q_{n+1}(z) - \zeta_{n+1} Q_n(z)}.$$

For $n > 0$, this transformation maps the axis of reals of the $\zeta_{n+1}$ plane onto the circumference of a certain circle $C_{n+1}(z)$ situated in the lower half-plane $\Im\zeta < 0$ in such a way that the points of the half-plane $\Im\zeta_{n+1} < 0$ are mapped onto the interior of this circle. The same is true if $n = 0$, with the exception that the circle $C_1(z)$ is tangent from below to the axis of reals at the origin. Comparing this transformation with formulas (2.38) and (2.39), we arrive at

THEOREM 2.7. *For a fixed z, $\Im z > 0$, the range of values which are assumed by solutions of problem $(N_n)$ is the closed circle $C_{n+1}(z)$. To each point on the circumference of $C_{n+1}(z)$ there corresponds a unique solution of $(N_n)$ of the type $\frac{p(z)}{q(z)}$, where $q(z)$ is a quasi-orthogonal polynomial of order $(n + 1)$. To each point interior to $C_{n+1}(z)$ there correspond infinitely many solutions.*

Indeed, a point on the circumference of $C_{n+1}(z)$ is of the type (2.38), where $\tau$ has a real value (including $\tau = \infty$), so that in this case $f_{n+1}(z) = \alpha_{n+1} + \tau$, and the corresponding solution $f(z)$ is uniquely determined. To a point $\zeta$ in the interior of $C_{n+1}(z)$ there correspond infinitely many solutions $f(z)$ — all those for which $f_{n+1}(z) - \alpha_{n+1} = \zeta_{n+1}$ — and it is obvious that there exist infinitely many functions $f_{n+1}(z)$ satisfying the above requirements and assuming at $z$ the given value $\alpha_{n+1} + \zeta_{n+1}$.

THEOREM 2.8. *The circles $\{C_n(z)\}$ of Theorem 2.7 decrease, as n increases, in such a way that $C_{n+1}(z)$ is inside $C_n(z)$ and touches $C_n(z)$ from inside. The radius $r_{n+1}(z)$ of $C_{n+1}(z)$ is given by*

$$r_{n+1}(z) = \frac{\rho_n(z)}{2y}, \qquad z = x + iy, \qquad y > 0. \tag{2.40}$$

The statement that $C_{n+1}(z)$ is inside $C_n(z)$ follows from the observation that the class of solutions of problem $(N_n)$ for any given $n$ is more inclusive than that for the next value $(n + 1)$. The two circles must have a point in common since the orthogonal polynomial of degree $n$ is a quasi-orthogonal polynomial of order $n$, as well as of order $(n + 1)$, and the corresponding point must lie simultaneously on the circumferences of $C_{n+1}(z)$ and of $C_n(z)$. To determine the radius of $C_{n+1}(z)$ it suffices to determine the radius of the circle described by $\zeta$, when $\zeta_{n+1}$ describes the axis of reals. The theory of complex linear transformations gives

$$r_{n+1}(z) = \frac{| P_{n+1}(z)Q_n(z) - Q_{n+1}(z)P_n(z) |}{2\Im[Q_{n+1}(z)\overline{Q_n(z)}]}.$$

The numerator of this fraction is equal to $\beta_0\beta_1 \cdots \beta_n$, by (2.37). As to the denominator, the recurrence relations yield

$$\mathfrak{J}[Q_{n+1}(z)\overline{Q_n(z)}] = y \mid Q_n(z) \mid^2 + \beta_n \mathfrak{J}[Q_n(z)\overline{Q_{n-1}(z)}] = \cdots$$

(2.41)
$$= y\left[ \mid Q_n(z) \mid^2 + \sum_{k=0}^{n-1} \beta_n \beta_{n-1} \cdots \beta_{k+1} \mid Q_k(z) \mid^2 \right],$$

so that finally

$$\frac{1}{2r_{n+1}(z)} = y \sum_{k=0}^{n} \frac{\mid Q_k(z) \mid^2}{\beta_0 \cdots \beta_k} = y \sum_{k=0}^{n} a_k^2 \mid Q_k(z) \mid^2$$
$$= y \sum_{k=0}^{n} \mid \omega_k(z) \mid^2 = y/\rho_n(z),$$

which is the desired result.

By Theorem 2.2, every solution $f(z)$ of problem $(N_n)$ can be represented as $f(z) = I(z;\psi_n)$, where $\psi_n(t)$ is a solution of the reduced moment problem (2.9), and conversely. Hence, the closed circle $C_{n+1}(z)$ represents the range of values of $I(z;\psi_n)$ for a fixed $z$, $\mathfrak{J}z > 0$, when $\psi_n$ ranges over all solutions of the reduced moment problem (2.9). As $n$ increases, $C_{n+1}(z)$ decreases and tends either to a point or to a limiting circle $C(z)$. The radius $r(z)$ of $C(z)$ is given by

(2.42)
$$r(z) = \frac{\rho(z)}{2y}, \qquad \frac{1}{\rho(z)} = \sum_{\nu=0}^{\infty} \mid \omega_\nu(z) \mid^2;$$

It reduces to 0 if and only if the series $\sum_{\nu=0}^{\infty} \mid \omega_\nu(z) \mid^2$ diverges.

On the other hand, it is readily seen that $C(z)$ is the range of values of $I(z;\psi)$, when $z$ is fixed, $\mathfrak{J}z > 0$, and $\psi(t)$ ranges over all solutions of the moment problem (2.1). Indeed, if $\psi(t)$ is such a solution, it is also a solution of all reduced moment problems for any $n$, so that $I(z;\psi)$ must be in $C(z)$. Now, any point $\zeta_0$ in $C(z)$ lies in all $C_{n+1}(z)$, so that for each $n$ we can find a solution $\psi^{(n)}(t)$ of the reduced moment problem of order $n$ such that $I(z;\psi^{(n)}) = \zeta_0$. On applying the Helly Theorems to the sequence $\{\psi^{(n)}(t)\}$, we can extract a subsequence $\{\psi^{(n_k)}(t)\}$ which tends substantially to a monotonic function $\psi(t)$. By Introduction, **3**, $\zeta_0 = I(z;\psi^{(n)}) \to I(z;\psi)$. It remains to prove that $\psi(t)$ is a solution of the moment problem (2.1). Since, for a fixed $j$ and sufficiently large $k$,

$$\int_{-\infty}^{\infty} t^j d\psi^{(n_k)}(t) = \mu_j,$$

it is sufficient to prove that

$$\int_{-\infty}^{\infty} t^j d\psi^{(n_k)}(t) \to \int_{-\infty}^{\infty} t^j d\psi(t), \quad \text{as } k \to \infty.$$

This, however, follows from Introduction, **3**, and from the fact that, for a fixed $j$, the integrals $\int_{-\infty}^{\infty} t^j d\psi^{(n_k)}(t)$ converge absolutely and uniformly in $k$. Indeed, if $2j_0$ is the smallest even number which is larger than $j$, we have

$$\int_T^\infty t^j d\psi^{(n_k)}(t) \leqq T^{j-2j_0} \int_T^\infty t^{2j_0} d\psi^{(n_k)}(t) \leqq \mu_{2j_0} T^{j-2j_0}, \qquad n_k \geqq j_0,$$

an analogous estimate being true also for $\int_{-\infty}^{-T} t^j d\psi^{(n_k)}(t)$.

Thus, $C(z)$ is the range of $I(z; \psi)$, when $\psi(t)$ ranges over all solutions of the moment problem (2.1). Since $I(z;\psi)$ is a continuous function of $z$ for all complex $z$, it is now clear that $C(z)$ depends continuously on $z$, which implies, in particular, that $\rho(z)$ is a continuous function of $z$ when $z$ is complex. It also follows that the condition that $C(z)$ reduces to a single point for all complex $z$, with $\mathfrak{I}z > 0$, is a necessary and sufficient condition that the moment problem (2.1) be determined. This condition will be satisfied if it holds for a sequence of values of $z$ which has a limiting point in the half-plane $y > 0$. But, by Theorem 2.6, $\rho(z)$ increases when $z$ approaches the axis of reals moving on a vertical line. Hence, the moment problem is determined if $\rho(z) = 0$ or, which is the same, $\sum_{\nu=0}^{\infty} | \omega_\nu(z) |^2 = \infty$, at a single point, real or complex. We may now state

THEOREM 2.9. *For each complex $z$ the circle $C(z) = \lim_n C_n(z)$ is the range of values of $I(z; \psi)$, when $\psi(t)$ ranges over all solutions of the moment problem* (2.1). *$C(z)$ depends continuously on $z$. In order that the moment problem* (2.1) *be determined it is necessary and sufficient that*

$$\rho(z) = 0, \quad or \quad \sum_{\nu=0}^{\infty} | \omega_\nu(z) |^2 = \infty \tag{2.43}$$

*at every complex point $z$. For this it is sufficient that* (2.43) *be satisfied at one point, real or complex.*

COROLLARY 2.7. *If the moment problem* (2.1) *is indeterminate, then, for all $z$,*

$$\rho(z) > 0, \quad or \quad \sum_{\nu=0}^{\infty} | \omega_\nu(z) |^2 < \infty.$$

**14.** Taking into consideration the generalized approximants $\mathcal{C}_{n+1}(z; \tau)$, we proceed to show that Theorem 2.9 can be restated in a different form, as first stated by Hamburger, [3].

DEFINITION 2.2. *The associated continued fraction*

$$\frac{\beta_0 |}{| z - \alpha_1} - \frac{\beta_1 |}{| z - \alpha_2} - \cdots \tag{2.44}$$

*is said to converge completely at a given point $z$ if, at this point, the sequence of its generalized approximants $\{\mathcal{C}_{n+1}(z; \tau)\}$ converges to the same limit for all real $\tau$ (including $\tau = \infty$) and uniformly in $\tau$.*

THEOREM 2.10. *A necessary and sufficient condition that the moment problem* (2.1) *be determined is that the associated continued fraction* (2.44) *shall converge completely for all complex z.*

Indeed, since $\mathcal{C}_{n+1}(z;\tau)$ describes the circumference of $C_{n+1}(z)$ when $\tau$ ranges from $-\infty$ to $+\infty$, it is clear that the complete convergence of the continued fraction (2.44) is equivalent to the condition that $C(z) = \underline{\lim}_n C_n(z)$ reduces to a single point.

COROLLARY 2.8. *If the moment problem* (2.1) *is determined, then* $\rho(z) = 0$ *at all complex z and also at all real z, except perhaps for at most a denumerable set of points.*

Only the second part of this corollary requires discussion. If the moment problem (2.1) is determined, then its solution $\psi(z)$ is substantially unique, and the point spectrum of $\psi(t)$ is at most a denumerable set. On the other hand, by Corollary 2.4, we can always construct a solution of our moment problem which has a concentrated mass $\rho(z)$ at any real $z$, and so, if the set of points where $\rho(z) > 0$ is not finite or denumerable, we arrive at a contradiction.

**15.** In the subsequent discussion we shall treat in more detail the case of an indeterminate moment problem, and hence assume that, in addition to the conditions of existence of a solution, $\Delta_n > 0$, $n = 0, 1, \cdots$, we also have $\rho(z) > 0$ for all $z$. Two sequences of generalized approximants $\mathcal{C}_{n+1}(z;\tau)$ of the continued fraction (2.44) are particularly valuable for our discussion. They are obtained by giving $\tau$ in (2.39) the values $\dfrac{P_{n+1}(0)}{P_n(0)}$ and $\dfrac{Q_{n+1}(0)}{Q_n(0)}$ respectively. Thus we obtain two sequences of functions

$$\mathcal{C}^{(1)}_{n+1}(z) = \frac{P_{n+1}(z)P_n(0) - P_n(z)P_{n+1}(0)}{Q_{n+1}(z)Q_n(0) - Q_n(z)Q_{n+1}(0)},$$

$$\mathcal{C}^{(2)}_{n+1}(z) = \frac{P_{n+1}(z)Q_n(0) - P_n(z)Q_{n+1}(0)}{Q_{n+1}(z)Q_n(0) - Q_n(z)Q_{n+1}(0)}.$$

Introducing four sequences of polynomials

$$(2.45)\qquad \begin{cases} A_{n+1}(z) = (\beta_0\beta_1\cdots\beta_n)^{-1}[P_{n+1}(z)P_n(0) - P_n(z)P_{n+1}(0)], \\ B_{n+1}(z) = (\beta_0\beta_1\cdots\beta_n)^{-1}[Q_{n+1}(z)P_n(0) - Q_n(z)P_{n+1}(0)], \\ C_{n+1}(z) = (\beta_0\beta_1\cdots\beta_n)^{-1}[P_{n+1}(z)Q_n(0) - P_n(z)Q_{n+1}(0)], \\ D_{n+1}(z) = (\beta_0\beta_1\cdots\beta_n)^{-1}[Q_{n+1}(z)Q_n(0) - Q_n(z)Q_{n+1}(0)], \end{cases}$$

we may write

$$\mathcal{C}^{(1)}_{n+1}(z) = \frac{A_{n+1}(z)}{B_{n+1}(z)}, \qquad \mathcal{C}^{(2)}_{n+1}(z) = \frac{C_{n+1}(z)}{D_{n+1}(z)}.$$

Let $\chi_n(z)$ be the numerator corresponding to the denominator $\omega_n(z)$, the $n$-th orthonormal polynomial. Using the recurrence relations, it is readily found that

$$(2.46)\quad \begin{cases} A_{n+1}(z) = z\sum_{\nu=0}^{n} P_\nu(0)P_\nu(z)(\beta_0\cdots\beta_\nu)^{-1} = z\sum_{\nu=0}^{n}\chi_\nu(0)\chi_\nu(z), \\ B_{n+1}(z) = -1 + z\sum_{\nu=0}^{n} P_\nu(0)Q_\nu(z)(\beta_0\cdots\beta_\nu)^{-1} = -1 + z\sum_{\nu=0}^{n}\chi_\nu(0)\omega_\nu(z), \\ C_{n+1}(z) = 1 + z\sum_{\nu=0}^{n} Q_\nu(0)P_\nu(z)(\beta_0\cdots\beta_\nu)^{-1} = 1 + z\sum_{\nu=0}^{n}\omega_\nu(0)\chi_\nu(z), \\ D_{n+1}(z) = z\sum_{\nu=0}^{n} Q_\nu(0)Q_\nu(z)(\beta_0\cdots\beta_\nu)^{-1} = z\sum_{\nu=0}^{n}\omega_\nu(0)\omega_\nu(z). \end{cases}$$

From (2.37) it is also immediately found that

$$(2.47)\quad A_{n+1}(z)D_{n+1}(z) - C_{n+1}(z)B_{n+1}(z) = 1.$$

A fundamental result is expressed now by

THEOREM 2.11. *If the moment problem* (2.1) *is indeterminate, that is, if* $\rho(z) > 0$ *for all* $z$, *then in the whole* $z$-*plane*

$$(2.48)\quad \frac{1}{\rho(z)} = e^{\delta(r)r}, \qquad |z| = r,$$

*where* $\delta(z)$ *is a generic notation for a function which is bounded and is* $o(1)$, *as* $r \to \infty$. *The polynomials* $A_{n+1}(z)$, $B_{n+1}(z)$, $C_{n+1}(z)$, $D_{n+1}(z)$ *converge, as* $n \to \infty$ *(uniformly in every bounded domain), to entire functions* $A(z)$, $B(z)$, $C(z)$, $D(z)$, *respectively. These entire functions are at most of order* 1 *and of minimal type, that is*

$$|A(z)|, |B(z)|, |C(z)|, |D(z)| = e^{\delta(r)r}.$$

*They also satisfy the relation*

$$(2.49)\quad A(z)D(z) - B(z)C(z) = 1.$$

*The series*

$$(2.50)\quad \frac{1}{\rho(z)} = \sum_{\nu=0}^{\infty} |\omega_\nu(z)|^2$$

*also converges uniformly in every finite domain and is a continuous function of* $z$.

Introduce the polynomial

$$F_{n+1}(z) = B_{n+1}(z)D_{n+1}(z).$$

The proof of Theorem 2.11 is based on two facts. We first establish the existence of positive constants $L_1$, $L_2$ such that, for any value of $z$, $\Im z = y > 0$,

$$(2.51)\quad \frac{y}{\mu_0} < \frac{y}{\rho_n(z)} \leq |F_{n+1}(z)| < \frac{L_1 r + L_2 r^2}{\rho_n(z)} < \frac{L_1 r + L_2 r^2}{\rho(z)}.$$

Using this we establish, in the second place, that

$$\int_{-\infty}^{\infty} \frac{\log \frac{\mu_0}{\rho(t)}}{1+t^2}\, dt < \infty. \tag{2.52}$$

The proof of Theorem 2.11 will be then readily concluded.

To prove (2.51) observe that the fractions $\frac{A_{n+1}(z)}{B_{n+1}(z)}$ and $\frac{C_{n+1}(z)}{D_{n+1}(z)}$ are generalized approximants of (2.44) and therefore their values lie on the circumference of the circle $C_{n+1}(z)$ of radius $r_{n+1}(z) = \frac{\rho_n(z)}{2y}$. The absolute value of the difference of these two fractions cannot exceed the diameter of $C_{n+1}(z)$, which yields

$$\left| \frac{A_{n+1}(z)}{B_{n+1}(z)} - \frac{C_{n+1}(z)}{D_{n+1}(z)} \right| = \frac{1}{| F_{n+1}(z) |} \leqq \frac{\rho_n(z)}{y}.$$

This, together with the obvious inequality $\rho_n(z) < \rho_0(z) = \mu_0$, establishes the first two inequalities (2.51). On the other hand, if we apply the Cauchy inequality to (2.46), we get respectively

$$| D_{n+1}(z) |^2 \leqq \frac{r^2}{\rho_n(0)\rho_n(z)} < \frac{r^2}{\rho(0)\rho_n(z)},$$

$$| B_{n+1}(z) |^2 \leqq 2\left[ 1 + \frac{\left( \sum_{\nu=0}^{n} | \chi_\nu(0) |^2 \right) r^2}{\rho_n(z)} \right]$$

Since all the roots of $\chi_\nu(z)$ are real, we have for any fixed $y_0 > 0$, $| \chi_\nu(0) | \leqq | \chi_\nu(iy_0) |$, while, by (2.23),

$$| \chi_\nu(iy_0) | \leqq \frac{\mu_0 | \omega_\nu(iy_0) |}{y_0}.$$

Thus we see that

$$\sum_{\nu=0}^{n} | \chi_\nu(0) |^2 \leqq \frac{\mu_0^2}{y_0^2} \sum_{\nu=0}^{n} | \omega_\nu(iy_0) |^2 < \frac{\mu_0^2}{y_0^2 \rho(iy_0)}. \tag{2.53}$$

On substituting this and estimating the product $| B_{n+1}(z) D_{n+1}(z) |$, we obtain the two remaining inequalities (2.51).

We now pass to the proof of (2.52). Take any value $y_0 > \mu_0$. Then, by (2.51),

$$\log | F_{n+1}(x + iy_0) | > \log \frac{y_0}{\rho_n(x + iy_0)} > \log \frac{y_0}{\mu_0} > 0.$$

Since all roots of $F_{n+1}(z)$ are real, it is readily seen by the Poisson formula [Titchmarsh, 2, p. 124], that

$$\log F_{n+1}(i + iy_0) = \frac{1}{\pi} \int_{-\infty}^{\infty} \frac{\log F_{n+1}(x + iy_0)}{1 + x^2}\, dx,$$

whence, comparing real parts,

$$\log | F_{n+1}(i + iy_0) | = \frac{1}{\pi} \int_{-\infty}^{\infty} \frac{\log | F_{n+1}(x + iy_0) |}{1 + x^2} dx.$$

But, by (2.51), the left-hand member here is bounded as $n \to \infty$. The same will be true for the right-hand member and, again by (2.51), for

$$\int_{-\infty}^{\infty} \log \frac{y_0}{\rho_n(x + iy_0)} \frac{dx}{1 + x^2}.$$

Since the integrand is an increasing function of $n$ we conclude that the integral

$$\int_{-\infty}^{\infty} \log \frac{\mu_0}{\rho(x + iy_0)} \frac{dx}{1 + x^2}$$

converges, and finally, since $\rho(x + iy)$ increases when $y \downarrow 0$, we see that this integral converges for all $y_0 \geqq 0$, and in particular, that

$$\int_{-\infty}^{\infty} \log \frac{\mu_0}{\rho(x)} \frac{dx}{1 + x^2} < \infty,$$

which is the desired result.

Now, for any real $z = x$, we see from (2.51) that

$$\log | F_{n+1}(x) | < \log \frac{\mu_0}{\rho(x)} + \log (1 + x^2) + L,$$

where $L$ is a suitable positive constant. Hence, by the Poisson formula, for any $z = x + iy$, $y > 0$, we have

$$\log | F_{n+1}(x + iy) | = \frac{y}{\pi} \int_{-\infty}^{\infty} \frac{\log | F_{n+1}(t) |}{(t - x)^2 + y^2} dt$$

$$< \frac{y}{\pi} \int_{-\infty}^{\infty} \frac{\log \frac{\mu_0}{\rho(t)} + \log (1 + t^2) + L}{(t - x)^2 + y^2} dt.$$

In the sector (2.5) we have, for $| t | \geqq 1$,

$$(t - x)^2 + y^2 \geqq \frac{t^2}{\sin^2 \epsilon} \geqq \frac{1 + t^2}{2 \sin^2 \epsilon} = \frac{1}{M} (1 + t^2).$$

Thus, for any $T \geqq 1$,

$$\log | F_{n+1}(x + iy) | \leqq \frac{1}{\pi y} \int_{-T}^{T} \left[ \log \frac{\mu_0}{\rho(t)} + \log (1 + t^2) + L \right] dt$$

$$+ \frac{yM}{\pi} \left( \int_{-\infty}^{-T} + \int_{T}^{\infty} \right) \frac{\log \frac{\mu_0}{\rho(t)} + \log (1 + t^2) + L}{1 + t^2} dt.$$

Given an arbitrarily small positive $\delta$, we first select $T$ (independent of $y$) so large that the second term is smaller than $\frac{y\delta}{2}$, and then $y$ so large that the first term is also less than $\frac{y\delta}{2}$. It follows that in the sector (2.5), for $|z| = r$ sufficiently large,

$$\log |F_{n+1}(z)| < \delta r,$$

and, in view of (2.51), also

$$\log \frac{1}{\rho_n(z)} < \delta r.$$

Using the fact that $\rho_n(z)$ decreases when $|y|$ increases and $x$ is fixed, we see that in either of the two sectors

$$0 \leqq \arg z \leqq \epsilon, \qquad \pi - \epsilon \leqq \arg z \leqq \pi,$$

we have, for $r$ sufficiently large,

$$\log \frac{1}{\rho_n(z)} < \frac{\delta r}{\cos \epsilon}.$$

By reflecting in the axis of reals we obtain the same inequalities in the lower half-plane $y \leqq 0$, and finally, on allowing $n \to \infty$, we see that, with no restriction on the location of $z$,

$$\log \frac{1}{\rho(z)} < \delta(r)r, \qquad \delta(r) = o(1), \quad \text{as} \quad r \to \infty,$$

$$\frac{1}{\rho(z)} = \sum_{\nu=0}^{\infty} |\omega_\nu(z)|^2 < e^{\delta(r)r}.$$

If, now, we observe that $\sum_{\nu=0}^{\infty} |\chi_\nu(0)|^2 < \infty$ and use Cauchy's inequality, we see at once that the limits

$$B(z) = \lim_n B_{n+1}(z) = -1 + z \sum_{\nu=0}^{\infty} \chi_\nu(0)\omega_\nu(z),$$

$$D(z) = \lim_n D_{n+1}(z) = z \sum_{\nu=0}^{\infty} \omega_\nu(0)\omega_\nu(z)$$

exist everywhere and are entire functions which admit of estimates analogous to (2.48). To establish the same properties for

$$A(z) = \lim_n A_{n+1}(z) = z \sum_{\nu=0}^{\infty} \chi_\nu(0)\chi_\nu(z),$$

$$C(z) = \lim_n C_{n+1}(z) = 1 + z \sum_{\nu=0}^{\infty} \omega_\nu(0)\chi_\nu(z),$$

it is sufficient to show that

$$\sum_{\nu=0}^{\infty} |\chi_\nu(z)|^2 < e^{\delta(r)r}.$$

This, however, follows from the fact that, for $\mathfrak{J}z = y \neq 0$,

$$|\chi_\nu(z)| < \frac{1}{|y|} |\omega_\nu(z)|,$$

so that

$$\sum_{\nu=0}^{\infty} |\chi_\nu(z)|^2 \leqq \frac{1}{y^2} \sum_{\nu=0}^{\infty} |\omega_\nu(z)|^2,$$

and also from the fact that the left-hand member of the preceding inequality is an increasing function of $|y|$, for a fixed $\mathfrak{R}z = x$. The uniform convergence of all these series is now obvious, while (2.49) is implied by (2.47). Finally, the continuity of $1/\rho(z)$ is implied by the continuity of the circle $C(z)$, and the uniform convergence of $\sum_{\nu=0}^{\infty} |\omega_\nu(z)|^2$ then follows from a well-known theorem of Dini.

The following result is easily derived from Theorem 2.11.

COROLLARY 2.9. *A necessary and sufficient condition that the moment problem* (2.1) *be determined is that the equation*

$$\underline{\mu}(F) = \bar{\mu}(F) \tag{2.54}$$

*hold for any fixed function $F(t)$ of the manifold $\mathfrak{M}_C$, distinct from a polynomial.* Thus,* (2.54) *holds either for all elements of $\mathfrak{M}_C$ or for none distinct from a polynomial.*

The necessity of (2.54) is trivial. To prove its sufficiency we show that if (2.1) is indeterminate, then (2.54) implies that $F(t)$ is a polynomial. Now, if (2.54) is satisfied, there exist two sequences of polynomials $\{P'_n(t)\}$, $\{P''_n(t)\}$, $n = 1, 2, \cdots$, such that

$$P'_n(t) \leqq F(t) \leqq P''_m(t), \qquad n, m = 1, 2, \cdots,$$

while

$$\mu(P'_n) \to \underline{\mu}(F), \qquad \mu(P''_m) \to \bar{\mu}(F),$$

so that

$$0 \leqq \mu(P''_m - P'_n) = \epsilon_{nm} \to 0, \qquad \text{as} \quad n, m \to \infty.$$

Using Corollary 2.1, we conclude that, for any $z$, real or complex,

$$|P''_m(z) - P'_n(z)| \leqq \frac{\epsilon_{nm}}{\rho(z)}.$$

Thus, both sequences $\{P'_n(z)\}$, $\{P''_n(z)\}$ converge to the same entire function $F(z)$ which coincides with $F(t)$ for real values of $z$ and which admits of estimates

$$|F(z)| < e^{\delta(r)r}, \qquad |z| = r,$$

$$|F(t)| = O(|t|^k), \qquad \text{as} \quad |t| \to \infty.$$

* For the notation used, see p. 10.

An easy application of the Phragmen-Lindelöf theorem [Titchmarsh, 2, p. 176] shows now that $F(z)$ is a polynomial.

**16.** We are now able to derive the general form $\psi(t)$ of any solution of the moment problem (2.1) in the indeterminate case, and the general form of the corresponding function $f(z) = I(z; \psi)$.

THEOREM 2.12. *If the moment problem* (2.1) *is indeterminate, then the general form of the function* $I(z; \psi)$ *is*

$$\int_{-\infty}^{\infty} \frac{d\psi(t)}{z - t} = I(z; \psi) = \frac{A(z) - \sigma(z)C(z)}{B(z) - \sigma(z)D(z)}, \tag{2.55}$$

*where* $A(z)$, $B(z)$, $C(z)$, $D(z)$ *are the entire functions of Theorem* 2.11 *and* $\sigma(z)$ *is an arbitrary function (including the case* $\sigma(z) = \infty$*), analytic in the half-plane* $y > 0$ *and satisfying the condition* $\mathfrak{I}\sigma(z) \leqq 0$, $\mathfrak{I}z > 0$. *To each solution* $\psi(t)$ *of the moment problem there corresponds just one function* $\sigma(z)$ *determined by* (2.55). *Conversely, to each function* $\sigma(z)$ *satisfying the above conditions there corresponds substantially one solution* $\psi(t)$ *of the moment problem given by the Stieltjes inversion formula. For each fixed non-real value of* $z$ *the function*

$$\frac{A(z) - \sigma C(z)}{B(z) - \sigma D(z)} \tag{2.56}$$

*describes the circumference of the circle* $C(z)$ *when the real parameter* $\sigma$ *describes* $[-\infty, \infty]$. *The value of* $I(z; \psi)$ *is always either on the boundary or in the interior of* $C(z)$. *To each point* $\zeta_0$ *of the circumference of a circle* $C(z_0)$ *there corresponds a substantially unique solution of the moment problem* $\psi(t)$ *such that* $I(z_0; \psi) = \zeta_0$. *It is given by the formula* (2.55), *with* $\sigma(z)$ *replaced by the constant* $\sigma_0$ *determined from*

$$\zeta_0 = \frac{A(z_0) - \sigma_0 C(z_0)}{B(z_0) - \sigma_0 D(z_0)}.$$

*The set of solutions which so correspond to all points of the circumference of* $C(z_0)$ *does not depend on* $z_0$. *To each point* $\zeta_0$ *interior to the circle* $C(z_0)$ *there correspond continuously many solutions* $\psi(t)$ *such that* $I(z; \psi) = \zeta_0$.

Assume that $\psi(t)$ is a solution of the moment problem (2.1). Then $f(z) = I(z; \psi)$ is a solution of problem $(N_n)$ for all values of $n$, and hence is represented by

$$f(z) = \frac{P_{n+1}(z) - [f_{n+1}(z) - \alpha_{n+1}]P_n(z)}{Q_{n+1}(z) - [f_{n+1}(z) - \alpha_{n+1}]Q_n(z)}, \tag{2.38}$$

where $f_{n+1}(z)$ is analytic in $y > 0$, $\mathfrak{I}f_{n+1}(z) \leqq 0$ and, in addition, $f_{n+1}(z) = o(z)$, as $z \to \infty$ in any sector (2.5). On introducing the function $\sigma_{n+1}(z)$ determined from

$$f_{n+1}(z) - \alpha_{n+1} = \frac{P_{n+1}(0) - Q_{n+1}(0)\sigma_{n+1}(z)}{P_n(0) - Q_n(0)\sigma_{n+1}(z)},$$

we can write (2.38) in the form

$$f(z) = \frac{A_{n+1}(z) - \sigma_{n+1}(z)C_{n+1}(z)}{B_{n+1}(z) - \sigma_{n+1}(z)D_{n+1}(z)}.$$

On allowing here $n \to \infty$, $\sigma_{n+1}(z) \to \sigma(z)$, where $\sigma(z)$ has the properties stated in the theorem, and

$$f(z) = \frac{A(z) - \sigma(z)C(z)}{B(z) - \sigma(z)D(z)}.$$

Conversely, take any function $f(z)$ of this form, where $\sigma(z)$ is analytic in $y > 0$, $\mathfrak{I}\sigma(z) \leqq 0$. For each $n$ put

$$f_{n+1}(z) - \alpha_{n+1} = \frac{P_{n+1}(0) - Q_{n+1}(0)\sigma(z)}{P_n(0) - Q_n(0)\sigma(z)}$$

and consider the function

$$f^{(n)}(z) = \frac{P_{n+1}(z) - [f_{n+1}(z) - \alpha_{n+1}]P_n(z)}{Q_{n+1}(z) - [f_{n+1}(z) - \alpha_{n+1}]Q_n(z)},$$

which can be written also as

$$f^{(n)}(z) = \frac{P_n(z) - [f_n(z) - \alpha_n]P_{n-1}(z)}{Q_n(z) - [f_n(z) - \alpha_n]Q_{n-1}(z)},$$

where

$$f_n(z) = \alpha_n + \frac{\beta_n}{z - f_{n+1}(z)}.$$

In view of the relation (2.6), it is seen that the transformation

$$\zeta = \frac{P_{n+1}(0) - Q_{n+1}(0)w}{P_n(0) - Q_n(0)w}$$

maps the axis of reals of the $w$-plane onto the axis of reals in the $\zeta$-plane, and the half-planes $\mathfrak{I}w \gtrless 0$ onto the half-planes $\mathfrak{I}\zeta \gtrless 0$ respectively. Hence, the function $f_{n+1}(z)$ is analytic in $y > 0$ and $\mathfrak{I}f_{n+1}(z) \leqq 0$. The same holds for $f_n(z)$ and, in addition, $f_n(z) = o(z)$, as $z \to \infty$ in the sector (2.5). It follows that $f^{(n)}(z)$ is a solution of problem $(N_n)$ for $n = 2, 3, \cdots$. But $f^{(n)}(z) \to f(z)$, as $n \to \infty$, and a repetition of the argument used on pp. 49–50 shows that $f(z) = I(z; \psi)$, where $\psi(t)$ is a solution of our moment problem. The statements of the theorem concerning the relationship of $\sigma(z)$ and $\psi(t)$ are clear from the previous discussion. The circle described by (2.56) is the limiting position of the circle

$$\frac{A_{n+1}(z) - \sigma C_{n+1}(z)}{B_{n+1}(z) - \sigma D_{n+1}(z)}, \qquad \sigma \text{ real},$$

which coincides with the circle

$$\frac{P_{n+1}(z) - \tau P_n(z)}{Q_{n+1}(z) - \tau Q_n(z)}, \qquad \tau \text{ real},$$

that is, with the circle $C_{n+1}(z)$ which, as we know, tends to $C(z)$, as $n \to \infty$. The statements concerning the case where $\zeta_0$ is on the boundary of the circle

$C(z_0)$ need no further discussion, while the final statement of Theorem 2.12 is implied by the fact that there exist continuously many distinct functions $\sigma(z)$ which satisfy the above requirements and assume a given value at a given point.

**17.** Some properties of the solutions $\psi(t)$ of the moment problem (2.1) are readily derived either directly or by analyzing formula (2.55). We mention here two properties referring for the proofs and also for some other properties to Stone [1].

1. Let $\Lambda$ be the closed set on the axis of reals consisting of points in every neighborhood of which lie roots of infinitely many ortho-normal polynomials $\omega_n(z)$. Then for any solution $\psi(t)$ of the moment problem (2.1)

$$\inf \mathfrak{S}(\psi) \leqq \inf \Lambda < \sup \Lambda \leqq \sup \mathfrak{S}(\psi),$$

while, in case the moment problem is determined,

$$\mathfrak{S}(\psi) \leqq \Lambda.$$

2. Let the moment problem (2.1) be indeterminate. The function $\sigma(z)$ of (2.55) can be written, by Lemma 2.1, as

$$\sigma(z) = Az + c + \int_{-\infty}^{\infty} \frac{1 + tz}{z - t}\, d\alpha(t),$$

where $\alpha(t)$ is a bounded increasing function, $A$ and $c$ are real constants, and $A \leqq 0$. With this representation the derived set of $\mathfrak{S}(\psi)$ coincides with the derived set of $\mathfrak{S}(\alpha)$. Hence, in the indeterminate case we can find solutions of the moment problem with spectrum, the derived set of which coincides with an arbitrarily given closed set.

The following criteria are due to Carleman [2].

1. The moment problem (2.1) is determined if the series

$$\sum_{n=0}^{\infty} \beta_n^{-\frac{1}{2}} \quad \text{diverges.}$$

This follows from the inequality

$$\beta_{n+1}^{-\frac{1}{2}} \leqq \tfrac{1}{2} y^{-1} \{ | \omega_n(z) |^2 + | \omega_{n+1}(z) |^2 \}, \qquad z = x + iy, \qquad y > 0,$$

which itself follows from (2.41).

2. If the continued fraction

$$\frac{\beta_0 |}{| z - \alpha_1} - \frac{\beta_1 |}{| z - \alpha_2} - \frac{\beta_2 |}{| z - \alpha_3} - \cdots$$

converges completely (that is, the corresponding moment problem is determined), then the following two continued fractions also converge completely:

$$\frac{\beta_0' \quad |}{| z - (\alpha_1 + \alpha_1')} - \frac{\beta_1' \quad |}{| z - (\alpha_2 + \alpha_2')} - \frac{\beta_2' \quad |}{| z - (\alpha_3 + \alpha_3')} - \cdots,$$

provided $\beta_\nu' > 0$ and the sequences $\{\alpha_\nu'\}$, $\{\beta_\nu^{1/2} - \beta_\nu'^{1/2}\}$ are bounded, and

$$\frac{\beta_0}{|z - \gamma_1 \alpha_1} \Big| - \frac{\beta_1}{|z - \gamma_2 \alpha_2} \Big| - \frac{\beta_2}{|z - \gamma_3 \alpha_3} \Big| - \cdots,$$

provided there exist two positive constants $l$, $L$ such that

$$l \leqq \gamma_\nu \leqq L, \qquad \nu = 1, 2, \cdots.$$

**18.** Assume that the moment problem (2.1) is indeterminate. Then Theorem 2.12 introduces a remarkable family of meromorphic functions (2.56) depending on the real parameter $\sigma$. In view of the relation (2.49), the functions corresponding to different values of $\sigma$ are distinct. Thus, to each value of $\sigma$ (including $\sigma = \pm\infty$) there corresponds a substantially uniquely determined solution of the moment problem (2.1) which we denote by $\psi_\sigma(t)$. These functions $\psi_\sigma(t)$ are called extremal solutions [Nevanlinna, 1] of the moment problem (2.1), in view of a certain extremal property which is stated in

THEOREM 2.13. *Let the moment problem* (2.1) *be indeterminate. Consider the family of meromorphic functions* (2.56) *and the corresponding solutions* $\{\psi_\sigma(t)\}$, *so that for each real value of* $\sigma$, *including* $\sigma = \pm\infty$,

$$\frac{A(z) - \sigma C(z)}{B(z) - \sigma D(z)} = I(z; \psi_\sigma) \equiv I_\sigma(z).$$

*Then* $\psi_\sigma(t)$ *is a step function; its spectrum coincides with the set of zeros of the denominator* $B(z) - \sigma D(z)$, *and the mass concentrated at each point* $x_j(\sigma)$ *of the spectrum precisely equals* $\rho(x_j(\sigma))$, *and hence is larger than the mass concentrated at this point by any other solution of the moment problem* (2.1).

*The numerator and the denominator of the meromorphic function* $I_\sigma(z)$ *have each infinitely many zeros which are all real and simple, and mutually separated; furthermore, the smallest (largest) root of the denominator, in case such exists, is smaller (larger) than the smallest (largest) root of the numerator. The roots of two denominators corresponding to two distinct values of* $\sigma$ *are also mutually separated. For any given real value* $x_0$ *there always exists a (unique) real value* $\sigma_0$ *(which may be infinite) such that the corresponding function* $\psi_0(t)$ *has* $x_0$ *in its spectrum with concentrated mass* $\rho(x_0)$. *The poles of* $I_\sigma(z)$, *or, which is the same, the points of the spectrum of* $\psi_\sigma(t)$, *are monotonic (analytic) functions of* $\sigma$. *If*

$$\cdots x_{-2}(-\infty), \quad x_{-1}(-\infty), \quad x_0(-\infty) = 0, \quad x_1(-\infty), \quad \cdots$$

*are the roots of the function* $D(z)$, *then, as* $\sigma$ *ranges over* $(-\infty, \infty)$ *the root* $x_j(\sigma)$ *of* $B(z) - \sigma D(z)$ *continuously increases from* $x_j(-\infty)$ *to* $x_{j+1}(-\infty)$.

For a proof we refer to Stone, [1]. Using Corollaries 2.3 and 2.4, the proof of Stone can be readily completed to yield the statements concerning the mass concentration at $x_j(\sigma)$ and $x_0$.

Theorem 2.13 shows that in the indeterminate case the moment problem (2.1) has always discontinuous extremal solutions, which form a family of solutions depending on a real parameter $\sigma$. By multiplying the extremal solution by a suitable factor depending on $\sigma$ and integrating with respect to $\sigma$ it is always possible to find infinitely many continuous, and even absolutely continuous, solutions of the moment problem [Hamburger, 3].

**19.** A remarkable property of extremal solutions of the moment problem (2.1) was discovered by M. Riesz, [4]. The present section will discuss this property. Let $\{\omega_n(z)\}$ be the sequence of ortho-normal polynomials determined by the sequence of moments $\{\mu_n\}$. Let $\psi(t)$ be any solution of the moment problem (2.1). We consider the class of complex-valued functions $f(t)$ of the real variable $t$ which are measurable with respect to $\psi(t)$ and for which

$$\int_{-\infty}^{\infty} |f(t)|^2 \, d\psi(t) < \infty. \tag{2.57}$$

Denote this class of functions by $L_\psi^2$ and the value of the integral (2.57) by $\|f\|_\psi^2$. Every function $f \in L_\psi^2$ generates the sequence of its Fourier coefficients relative to $\{\omega_n(z)\}$,

$$f_n = \int_{-\infty}^{\infty} f(t)\omega_n(t) \, d\psi(t), \qquad n = 0, 1, 2, \cdots,$$

and the corresponding formal Fourier series

$$\sum_{\nu=0}^{\infty} f_\nu \omega_\nu(t),$$

the $(n + 1)$-th partial sum of which will be denoted by

$$s_n(t; f) \equiv s_n(t) \equiv s_n(f) = \sum_{\nu=0}^{n} f_\nu \omega_\nu(t).$$

It is well known (Bessel's inequality) that

$$\| s_n(f) \|_\psi^2 = \sum_{\nu=0}^{n} |f_\nu|^2 \leq \int_{-\infty}^{\infty} |f(t)|^2 \, d\psi(t),$$

and that for any polynomial $P_n$, of degree $\leqq n$,

$$\|f - s_n\|_\psi^2 = \|f - P_n\|_\psi^2 + \|P_n - s_n\|_\psi^2,$$

so that

$$\|f - s_n\|_\psi^2 = \int_{-\infty}^{\infty} |f|^2 \, d\psi - \sum_{\nu=0}^{n} |f_\nu|^2 = \min_{P_n} \|f - P_n\|_\psi^2 .$$

The problem now is to find conditions on $\psi$ under which this minimum tends to 0, as $n \to \infty$, for every function $f \in L_\psi^2$, or, in other words, to find conditions under which the *Parseval formula*

$$\sum_{\nu=0}^{\infty} |f_\nu|^2 = \int_{-\infty}^{\infty} |f(t)|^2 \, d\psi(t)$$

holds for all functions $f \in L_\psi^2$. The solution of this problem is given by

THEOREM 2.14. *A necessary and sufficient condition that the Parseval formula hold for every function $f \in L_\psi^2$ is that either the moment problem* (2.1) *be determined, or, if it is indeterminate, that $\psi(t)$ be an extremal solution.*

We first show that the class $L_\psi^2$ can be replaced by the much more restricted subclass of functions of the type $\frac{1}{t - z_0}$, where $z_0$ is any non-real number. This will follow from

LEMMA 2.14. *The linear manifold $\mathfrak{M}\left(\frac{1}{t - z_0}\right)$ determined by the functions $\frac{1}{t - z_0}$ is dense in $L_\psi^2$.*

Assume that Lemma 2.14 is proved. It is trivial that the validity of the Parseval formula for all functions of $L_\psi^2$ implies its validity for all functions $\frac{1}{t - z_0}$. Now assume that the Parseval formula holds for all functions $\frac{1}{t - z_0}$, with $z_0$ non-real, and prove that it holds then for the whole class $L_\psi^2$. First, it is clear that if the Parseval formula holds for any function $\frac{1}{t - z_0}$, it will also hold for every element of the manifold $\mathfrak{M}\left(\frac{1}{t - z_0}\right)$. If now $\mathfrak{M}\left(\frac{1}{t - z_0}\right)$ is dense in $L_\psi^2$, given $\epsilon > 0$, we can find a $g_\epsilon(t) \in \mathfrak{M}\left(\frac{1}{t - z_0}\right)$ such that

$$\|f - g_\epsilon\|_\psi < \frac{\epsilon}{3}.$$

Furthermore,

$$\|f - s_n(f)\|_\psi \leqq \|f - g_\epsilon\|_\psi + \|g_\epsilon - s_n(g_\epsilon)\|_\psi + \|s_n(g_\epsilon) - s_n(f)\|_\psi .$$

Here the first term on the right is less than $\frac{\epsilon}{3}$, and the same is true of the third term, in view of

$$\|s_n(g_\epsilon) - s_n(f)\|_\psi^2 = \|s_n(g_\epsilon - f)\|_\psi^2 \leqq \|g_\epsilon - f\|_\psi^2 .$$

The second term on the right can be made less than $\frac{\epsilon}{3}$, if $n$ is taken sufficiently large, since we have assumed that the Parseval formula holds for $g_\epsilon \in \mathfrak{M}\left(\frac{1}{t - z_0}\right)$. Thus

$$\|f - s_n(f)\|_\psi < \epsilon,$$

for $n$ sufficiently large, which establishes the Parseval formula for the general element $f \in L_\psi^2$.

We now pass on to the proof of Lemma 2.14. A general element of $L_\psi^2$ can be approximated arbitrarily closely in $L_\psi^2$ by a function vanishing outside a finite interval and being continuous together with its first derivative. Hence, it suffices to consider only such elements of $L_\psi^2$. Let $g(t)$ be such an element. Consider the two functions

$$g_1(t) = g(t) + g(-t), \qquad g_2(t) = \frac{1 + t^2}{2t}[g(t) - g(-t)].$$

They are both continuous, vanish outside a finite interval, and depend only on $t^2$. It is easy to show that functions of this type can be approximated uniformly in $(-\infty, \infty)$ by linear combinations of fractions $\frac{1}{t^2 + b^2}$, $b \neq 0$ and real. This leads to a uniform approximation of $g(t)$ by a linear combination of fractions

$$\frac{1}{t^2 + b^2} = \frac{1}{2bi}\left(\frac{1}{t - ib} - \frac{1}{t + ib}\right)$$

and

$$\frac{t}{t^2 + b^2} = \frac{1}{2}\left(\frac{1}{t - ib} + \frac{1}{t + ib}\right),$$

which are of the desired type.

Now let $g(t^2)$ be a function which is continuous and vanishes outside a finite interval $(-a^{\frac{1}{2}}, a^{\frac{1}{2}})$. On setting $t^2 = x$ we obtain a function $g(x)$ which is defined for $x \geqq 0$, is continuous, and vanishes for $x > a$. Since $(0, \infty)$ is transformed into $(0, a)$ by the transformation $u = \frac{1}{x + a}$, functions of the above type can be approximated uniformly in $(0, \infty)$ by polynomials in $\frac{1}{x + a}$. Since, however, $\frac{1}{(x + a)^k}$ can be approximated uniformly by

$$1\Big/\left[\prod_{i=1}^{n}(x + a_i)\right] = \sum_{i=1}^{n} \frac{A_i}{x + a_i}, \qquad a_i \neq a_j, \qquad a_i > 0,$$

every polynomial in $\frac{1}{x + a}$ can be uniformly approximated by a linear combination of fractions of the type $\frac{1}{x + a_i}$, which, for a function of the type $g(t^2)$ above, gives a uniform approximation by fractions of the type $\frac{1}{t^2 + b^2}$, $b \neq 0$ and real. This proves our assertion and also completes the proof of Lemma 2.14.

**20.** We have proved that a necessary and sufficient condition that the Parseval formula hold for any function $f \in L_\psi^2$ is that it holds for any function of the type $1/(t - z_0)$, $z_0$ non-real.

Now let $z_0$ be any fixed non-real number, and let

$$s_n(t) \equiv s_n\left(t, \frac{1}{t - z_0}\right).$$

We wish to find necessary and sufficient conditions in order that

$$\rho_n^*(z_0) \equiv \int_{-\infty}^{\infty} \left| \frac{1}{t - z_0} - s_n(t) \right|^2 d\psi(t)$$

tends to 0, as $n \to \infty$. Observe that

$$\rho_n^*(z_0) = \min_{p_n} \int_{-\infty}^{\infty} \left| \frac{1}{t - z_0} - p_n(t) \right|^2 d\psi(t),$$

where the minimum is taken over all polynomials $p_n(t)$, of degree $\leqq n$. Inasmuch as

$$P_n(t) = 1 - (t - z_0)p_n(t)$$

can be considered as an arbitrary polynomial of degree $\leqq n + 1$ which assumes the value 1 at $t = z_0$, the minimum problem which determines $\rho_n^*(z_0)$ coincides with that treated on pp. 39–41, where $n$ is replaced by $(n + 1)$ and the distribution $d\psi(t)$ is replaced by $\dfrac{d\psi(t)}{|t - z_0|^2}$. Thus we may say that $\rho_n^*(z_0) \to \rho^*(z_0)$, as $n \to \infty$, and that the condition $\rho^*(z_0) = 0$ is necessary and sufficient that the moment problem with the distribution $\dfrac{d\psi(t)}{|t - z_0|^2}$ be determined. Thus, a necessary and sufficient condition that the Parseval formula hold for every function $f \in L_\psi^2$ is that the moment problem with a distribution of masses given by $\dfrac{d\psi(t)}{|t - z_0|^2}$ be determined for every fixed non-real $z_0$.

This condition, in turn, is equivalent to the following one: an increasing bounded function $\psi'(t)$ satisfying the conditions

$$\int_{-\infty}^{\infty} \frac{t^n\, d\psi(t)}{|t - z_0|^2} = \int_{-\infty}^{\infty} \frac{t^n\, d\psi'(t)}{|t - z_0|^2}, \qquad n = 0, 1, 2, \cdots, \tag{2.58}$$

must be substantially equal to $\psi(t)$. Conditions (2.58), however, are equivalent to

$$\int_{-\infty}^{\infty} t^n\, d\psi(t) = \int_{-\infty}^{\infty} t^n\, d\psi'(t), \qquad n = 0, 1, 2, \cdots, \tag{2.59}$$

with the additional condition

$$\int_{-\infty}^{\infty} \frac{d\psi(t)}{z_0 - t} = \int_{-\infty}^{\infty} \frac{d\psi'(t)}{z_0 - t}. \tag{2.60}$$

Indeed, if we write (2.60) in the form

$$\int_{-\infty}^{\infty} (\bar{z}_0 - t) \frac{d\psi(t)}{|z_0 - t|^2} = \int_{-\infty}^{\infty} (\bar{z}_0 - t) \frac{d\psi'(t)}{|z_0 - t|^2}$$

and observe that

$$t^n = \frac{(z_0 - t)(\bar{z}_0 - t)t^n}{|z_0 - t|^2},$$

it is immediately seen that (2.58) implies (2.59) and (2.60). Conversely, let (2.59) and (2.60) be satisfied. Let $z_0 = x_0 + iy_0$, $y_0 \neq 0$. By separating real and imaginary parts in (2.60), we easily prove (2.58) for $n = 0, 1$. To prove the conditions (2.58) for $n = 2, 3, \cdots$, it is sufficient to observe that

$$| z_0 - t |^2 = (x_0 - t)^2 + y_0^2$$

is a polynomial of the second degree in $t$, so that

$$t^n = | z_0 - t |^2 P_{n-2}(t) + A_n t + B_n,$$

where $P_{n-2}(t)$ is a polynomial of degree $(n - 2)$ and $A_n$ and $B_n$ are constants.

Thus, our necessary and sufficient conditions are equivalent to the statement that, given an arbitrary non-real $z_0$, whenever we have a solution $\psi'(t)$ of the moment problem (2.1) which satisfies, in addition, condition (2.60), then $\psi'(t)$ is substantially equal to $\psi(t)$. This means, however, that either problem (2.1) is determined, or, in case it is not, $I(z_0 ; \psi)$ lies on the boundary of the circle $C(z_0)$ (there is clearly no loss of generality in assuming that $\mathfrak{I} z_0 > 0$), which implies that $\psi(t)$ is an extremal solution. The proof of Theorem 2.14 is now complete.

Suppose that the moment-problem (2.1) is indeterminate, that $\psi(t)$ is a solution of (2.1) and that $f(t)$ is any fixed element of the class $L_\psi^2$. If $\{f_n\}$ are the Fourier coefficients of $f(t)$ relative to the ortho-normal set $\{\omega_n(z)\}$, the series $\sum_{n=0}^{\infty} | f_n |^2$ converges, and in view of Corollary 2.7 and Theorem 2.11, it is clear that the series $\sum_{n=0}^{\infty} f_n \omega_n(z)$ converges in $L_\psi^2$ on the axis of reals, and also uniformly in every bounded domain of the complex $z$-plane to an entire function $f^*(z)$, which admits of an estimate

$$f^*(z) = e^{\delta(r) r}, \qquad | z | = r, \qquad \delta(r) = o(1), \qquad \text{as} \quad r \to \infty.$$

The function $f^*(z)$ will have the same Fourier coefficients as $f(z)$, and it is easily seen that

$$\int_{-\infty}^{\infty} | f(t) |^2 \, d\psi(t) = \sum_{n=0}^{\infty} | f_n |^2 + \int_{-\infty}^{\infty} | f - f^* |^2 \, d\psi(t).$$

It follows that *a necessary and sufficient condition that the Parseval formula hold for a given function $f(t)$ is that $f(t)$ be equal to the sum of its Fourier series, $f^*(t)$, except perhaps at a set of points of $\psi$-measure* 0.

It should also be observed that functions of the type $f^*(t)$, or, which is the same, the sums of series $\sum_0^{\infty} \alpha_n \omega_n(t)$, where $\sum_{n=0}^{\infty} | \alpha_n |^2 < \infty$, are such that the value of $\int_{-\infty}^{\infty} f^*(t) \, d\psi(t)$ does not depend on the choice of the solution $\psi(t)$ of (2.1). In Chapter IV we shall see other examples of functions possessing this property. A complete determination of the class of functions which have this property is still an unsolved problem.

In discussing the validity of the Parseval formula in the Stieltjes case we may use arguments analogous to and even simpler than those used in the Hamburger case, with the conclusion that *a necessary and sufficient condition that the Parseval formula holds in the Stieltjes case is that the distribution* $\frac{d\psi(t)}{(t-x)^2}$ *correspond to a determinate Hamburger moment problem for any* $x < 0$.

Thus, in the Stieltjes case, the Parseval formula holds or does not hold according as

$$\lim_{n} \min_{P_n} \int_0^\infty (P_{n+1}(t))^2 \frac{d\psi(t)}{(t-x)^2} = 0 \quad \text{or} \quad > 0,$$

where the minimum is taken over all polynomials $P_{n+1}(t)$, of degree $\leqq n + 1$, which assume the value 1 at $t = x$, for all negative $x$.

On observing that, for such $x$ and $0 \leqq t < \infty$, $(t - x)^2$ is greater than each of the quantities $(-x)^2$, $t(-x)$, $t^2$, $(-x)(t - x)$ it is obvious that, in the Stieltjes case, the Parseval formula holds whenever one of the following distributions

$$d\psi(t), \qquad \frac{d\psi(t)}{t}, \qquad \frac{d\psi(t)}{t^2}, \qquad \frac{d\psi(t)}{t-x}$$

corresponds to a determined Hamburger moment problem.

**21.** We have already established necessary and sufficient conditions that the moment problem (2.1) be determined or indeterminate. Various other criteria were established by Hamburger, [3] and proved in a simpler way by Nevanlinna [1] and M. Riesz [1, 2, 3]. Here our exposition follows in general the line of argument of M. Riesz. We shall need a few more properties of quasi-orthogonal polynomials.

As was stated in Lemma 2.7, all quasi-orthogonal polynomials of order $(n + 1)$ which have a common root $x_0$ differ only by a constant factor. A quasi-orthogonal polynomial of order $(n + 1)$ which has a given root $x_0$ can be immediately given by

$$q(z) = \frac{1}{\Delta_n} \left| z^i \, x_0^i \, \mu_1 \cdots \mu_{i+n-1} \right|_0^{n+1}.$$

This is readily proved by verifying the condition $q(x_0) = 0$ and the conditions of quasi-orthogonality

$$\mu^i q(\mu) = 0, \qquad i = 0, 1, \cdots, n - 1.$$

In particular, a quasi-orthogonal polynomial of order $(n + 1)$ which vanishes at the origin is, up to a constant factor, given by

$$q(z) = \frac{1}{\Delta_n} \left| z^i \, \mu_i \, \mu_{i+1} \cdots \mu_{i+n-1} \right|_1^{n+1}. \tag{2.61}$$

Hence, the polynomial $D_{n+1}(z)$ in (2.46) can differ only by a constant factor from the above expression. We prove that

$$D_{n+1}(z) = \frac{1}{\Delta_n} \left| z^i \, \mu_i \, \mu_{i+1} \cdots \mu_{i+n-1} \right|_1^{n+1}.$$

It suffices to verify that the values of the corresponding numerators at a single point coincide. Now, formulas (2.45) show that the numerator corresponding to $D_{n+1}(z)$ is $C_{n+1}(z)$ and that (see (2.46))

$$C_{n+1}(0) = 1.$$

The numerator corresponding to the polynomial (2.61) is obtained if we replace in (2.61) $z^i$ by the polynomial

$$\frac{z^i - \mu^i}{z - \mu} = z^{i-1}\mu_0 + z^{i-2}\mu_1 + \cdots + z\mu_{i-2} + \mu_{i-1}.$$

This shows that if $p(z)$ is the numerator corresponding to $q(z)$, then $p(0) = 1 = C_{n+1}(0)$. Thus, $D_{n+1}(z) \equiv q(z)$.

The polynomial $B_{n+1}(z)$ of (2.45) assumes the value $(-1)$ at $z = 0$. Hence, the general quasi-orthogonal polynomial $q(z)$ of order $(n + 1)$ which assumes the value $(-1)$ at $z = 0$ can be represented by $B_{n+1}(z) + \mu_{-1}D_{n+1}(z)$, where $\mu_{-1}$ is an arbitrary parameter. To determine the value of $\mu_{-1}$ it is sufficient to prescribe the value of the numerator $p(z)$ corresponding to $q(z)$ at a given point, for instance, at the origin. By (2.45),

$$p(z) = A_{n+1}(z) + \mu_{-1}C_{n+1}(z), \qquad p(0) = \mu_{-1}.$$

On the other hand, the polynomial

$$-\frac{1}{\Delta_n} \left| z^i \mu_{i-1} \mu_i \cdots \mu_{i+n-1} \right|_0^{n+1}$$

is clearly quasi-orthogonal of order $(n + 1)$, assumes the value $(-1)$ at $z = 0$ and depends on the parameter $\mu_{-1}$. Moreover, by the above method, it is readily found that the numerator corresponding to this polynomial assumes precisely the value $\mu_{-1}$ at $z = 0$. This proves that

$$Q_{n+1}(z; \mu_{-1}) \equiv B_{n+1}(z) + \mu_{-1} D_{n+1}(z) = -\frac{1}{\Delta_n} \left| z^i \mu_{i-1} \mu_i \cdots \mu_{i+n-1} \right|_0^{n+1};$$

in particular,

$$B_{n+1}(z) = -\frac{1}{\Delta_n} \begin{vmatrix} 1 & 0 & \mu_0 & \mu_1 & \cdots & \mu_{n-1} \\ z & \mu_0 & \mu_1 & \mu_2 & \cdots & \mu_n \\ \cdots & \cdots & \cdots & \cdots & \cdots & \cdots \\ z^{n+1} & \mu_n & \mu_{n+1} & \mu_{n+2} & \cdots & \mu_{2n} \end{vmatrix}.$$

Now, on setting

$$W_n(z; \mu_{-1}) \equiv z^{-1}(B_{n+1}(z) + 1) + \mu_{-1}\frac{D_{n+1}(z)}{z} = \frac{1}{z}[Q_{n+1}(z; \mu_{-1}) + 1],$$

we see from (2.46), (2.45) that the Fourier coefficients of $W_n(z; \mu_{-1})$ relative to the orthonormal set $\{\omega_\nu(z)\}$ are $\chi_\nu(0) + \mu_{-1}\omega_\nu(z)$. Hence,

$$\delta_n \equiv \delta_n(\mu_{-1}) = W_n^2(\mu; \mu_{-1}) = \sum_{\nu=0}^{n} [\chi_\nu(0) + \mu_{-1}\omega_\nu(0)]^2.$$

The sequence $\{\delta_n\}$ is clearly increasing.

Let $x_0, x_1, \cdots, x_n$ be the roots of $Q_{n+1}(z; \mu_{-1})$ and let $\psi_n(t; \mu_{-1})$ be the corresponding distribution function of order $n$, so that the set $\{x_0, x_1, \cdots, x_n\}$ is the spectrum of $\psi_n(t; \mu_{-1})$. Then

$$\frac{P_{n+1}(z; \mu_{-1})}{Q_{n+1}(z; \mu_{-1})} = \int_{-\infty}^{\infty} \frac{d\psi_n(t; \mu_{-1})}{z - t} = \sum_{j=0}^{n} \frac{\rho_n(x_j)}{z - x_j},$$

and since $Q_{n+1}(0; \mu_{-1}) = -1$, while $P_{n+1}(0; \mu_{-1}) = \mu_{-1}$, we have

$$\mu_{-1} = \sum_{j=0}^{n} \frac{\rho_n(x_j)}{x_j} = \int_{-\infty}^{\infty} \frac{d\psi_n(t; \mu_{-1})}{t} \tag{2.62}$$

On the other hand, on applying the quadrature formula (2.21) to the polynomial $W_n^2(z; \mu_{-1})$ which assumes the value $\frac{1}{x_j^2}$ at the point $x_j$, we get

$$\delta_n = \sum_{j=0}^{n} \frac{\rho_n(x_j)}{x_j^2} \tag{2.63}$$

**22.** We are now prepared to prove the following three theorems.

THEOREM 2.15. *A necessary and sufficient condition that the moment problem* (2.1) *be determined is that at least one of the two sequences* $\{1/\rho_n(0)\}$ *and* $\{\delta_n(\mu_{-1})\}$, *for some value of the parameter* $\mu_{-1}$, *should diverge to* $\infty$

THEOREM 2.16. *A necessary and sufficient condition that the moment problem* (2.1) *be indeterminate is that, for two distinct values* $\mu_{-1}'$, $\mu_{-1}''$ *of the parameter* $\mu_{-1}$, *the sequences* $\{\delta_n(\mu_{-1}')\}$, $\{\delta_n(\mu_{-1}'')\}$ *should converge.*

THEOREM 2.17. *A necessary and sufficient condition that the moment problem* (2.1) *be determined is that at least one of the two series*

$$\sum_{\nu=0}^{\infty} \omega_\nu(0)^2, \qquad \sum_{\nu=0}^{\infty} \chi_\nu(0)^2 \tag{2.64}$$

*should diverge.*

To prove Theorem 2.15 we observe first that if $1/\rho_n(0) \to \infty$ or $\rho_n(0) \to 0$, then the moment problem (2.1) is determined. Therefore we assume that $1/\rho_n(0)$ converges to a finite limit and prove that, if for some value of $\mu_{-1}$ $\delta_n(\mu_{-1})$ also converges to a finite limit, or, which is the same, remains bounded, then the problem (2.1) is indeterminate. We know that if $\rho(0) = \lim_n \rho_n(0) \neq 0$, there exists a solution $\psi(t)$ of the moment problem (2.1) with the mass $\rho(0)$ concentrated at $t = 0$. On the other hand, for a given $\epsilon > 0$, we have

$$\psi_n(\epsilon) - \psi_n(-\epsilon) \leqq \epsilon^2 \sum_{|x_j| \leqq \epsilon} \rho_n(x_j) x_j^{-2} \leqq \epsilon^2 \delta, \tag{2.65}$$

where $\delta = \lim_n \delta_n(\mu_{-1}) < \infty$. By the Helly Theorem, there exists a subsequence $\{\psi_{n_k}(t)\}$ which tends substantially to a solution $\psi'(t)$ of the moment problem (2.1). If now $(-\epsilon, \epsilon)$ is an interval of continuity of $\psi'(t)$, then we must have

$$\psi'(\epsilon) - \psi'(-\epsilon) \leqq \epsilon^2 \delta,$$

which shows that $\psi'(t)$ is continuous at $t = 0$, and hence distinct from $\psi(t)$. Thus, the moment-problem (2.1) is indeterminate.

If, however, $\delta_n(\mu_{-1}) \to \infty$, then the moment problem (2.1) is determined since we know **[15]** that in the indeterminate case both series

$$\sum_{\nu=0}^{\infty} \omega_\nu(0)^2, \qquad \sum_{\nu=0}^{\infty} \chi_\nu(0)^2$$

converge, and therefore

$$\delta = \lim_n \delta_n(\mu_{-1}) = \sum_{\nu=0}^{\infty} [\chi_\nu(0) + \mu_{-1}\omega_\nu(0)]^2 < \infty.$$

To prove Theorem 2.16 we observe that if, for any value of $\mu_{-1}$, $\delta_n(\mu_{-1}) \to \infty$, then the problem (2.1) is determined, as we have just seen. Assume now that $\delta_n(\mu_{-1}) \to \delta(\mu_{-1}) < \infty$, for two distinct values $\mu_{-1} = \mu_{-1}', \mu_{-1}''$. This clearly is equivalent to the assumption that both series (2.64) converge. Then the argument used in connection with (2.65) shows the existence of two solutions $\psi'(t)$, $\psi''(t)$ of the moment problem (2.1), continuous at $t = 0$, which are limit functions of two subsequences extracted from $\{\psi_n(t, \mu_{-1}')\}$, $\{\psi_n(t, \mu_{-1}'')\}$ respectively. It remains to show that $\psi'(t)$ and $\psi''(t)$ are distinct. This will be done if we prove the possibility of passing to the limit in the relation (2.62), where $\mu_{-1}$ is equal to either of the two values $\mu_{-1}'$, $\mu_{-1}''$. Indeed, this will show that

$$\mu_{-1}' = \int_{-\infty}^{\infty} \frac{d\psi'(t)}{t} \neq \mu_{-1}'' = \int_{-\infty}^{\infty} \frac{d\psi''(t)}{t}.$$

Let $(-\epsilon, \epsilon)$ be a common interval of continuity of the functions $\psi'(t)$, $\psi''(t)$, arbitrarily small but fixed. It is obviously possible to pass to the limit under the sign of both integrals $\int_{-\infty}^{-\epsilon}$, $\int_{\epsilon}^{\infty}$, and it remains to investigate $\int_{-\epsilon}^{\epsilon}$. By Schwarz' inequality, we have

$$\int_{-\epsilon}^{\epsilon} \frac{d\psi_n(t, \mu_{-1})}{t} \leqq [\psi_n(\epsilon, \mu_{-1}) - \psi_n(-\epsilon, \mu_{-1})]^{\frac{1}{2}} \left[\int_{-\epsilon}^{\epsilon} \frac{d\psi_n(t, \mu_{-1})}{t^2}\right]^{\frac{1}{2}} < \epsilon\delta(\mu_{-1}),$$

so that the contribution of the interval $(-\epsilon, \epsilon)$ is uniformly $O(\epsilon)$, which justifies the passing to the limit in (2.62). This completes the proof of Theorem 2.16. The proof of Theorem 2.17 is also contained in the previous argument.

Another criterion was initially proved by Hamburger [3], and a simpler proof of it was given by M. Riesz, [1]. We consider, together with $\rho_n(0)$ and $\rho(0) = \lim_n \rho_n(0)$, the quantities $\rho_n^{(2)}(0)$, $\rho^{(2)}(0) = \lim_n \rho_n^{(2)}(0)$, which are determined by the sequence of moments $(\mu_2, \mu_3, \cdots)$ in the same way as $\rho_n(0)$, $\rho(0)$ are determined by the sequence $(\mu_0, \mu_1, \mu_2, \cdots)$. Hamburger's criterion is expressed by the following theorem.

THEOREM 2.18. *A necessary and sufficient condition that the moment problem* (2.1) *be determined is that at least one of the quantities*

$$\rho(0) = \lim_n \rho_n(0), \qquad \rho^{(2)}(0) = \lim_n \rho_n^{(2)}(0)$$

*be equal to zero.*

Indeed, if $\rho(0) = 0$, we know that (2.1) is determined. If $\rho^{(2)}(0) = 0$, then the moment-problem corresponding to the moments $(\mu_2, \mu_3, \cdots)$ must be determined, and therefore (2.1) must also be determined. This proves the necessity of our condition. The sufficiency follows from the important inequality

$$(2.66) \qquad 1/\rho(t) \leqq e^{4A^2t^2+6A|t|}(1/\rho(0) + t^2/\rho^{(2)}(0)),$$

where $t$ is any real number and

$$A = \left(\frac{\mu_0 - \rho(0)}{\rho^{(2)}(0)}\right)^{\frac{1}{2}} > 0.$$

Indeed, this inequality shows that if $\rho(0) > 0$, $\rho^{(2)}(0) > 0$, then $\rho(t) > 0$ and hence $\rho(z) > 0$ for all $z$, so that the problem (2.1) is indeterminate. We refer to M. Riesz [1] for an elegant proof of (2.66). M. Riesz [2] gave also a direct proof of the equivalence of conditions of Theorems 2.15 and 2.18.

**23.** In Theorem 2.5 $\rho_n(z_0)$ was defined as

$$\min_{P_n} \overline{P}_n(\mu)P_n(\mu)$$

for all polynomials $P_n(z)$, of degre $\leqq n$, which assume the value 1 at $z = z_0$. On setting $z_0 = 0$ and

$$P_n(z) = x_0 + x_1 z + \cdots + x_n z^n,$$

we see that $\rho_n(0)$ is the minimum of the quadratic form

$$\mathfrak{F}_n(x_0, x_1, \cdots, x_n) \equiv \sum_{i,j=0}^{n} \mu_{i+j} x_i x_j$$

on the plane $x_0 = 1$. Hamburger [3] introduces the formal quadratic form in infinitely many variables

$$\mathfrak{F}(x_0, x_1, \cdots) = \sum_{i,j=0}^{\infty} \mu_{i+j} x_i x_j$$

all of whose sections $\mathfrak{F}_n \equiv \mathfrak{F}_n(x_0, x_1, \cdots, x_n)$ are positive definite, in view of our basic condition $\Delta_n > 0$, $n = 0, 1, \cdots$, and calls $\mathfrak{F}(x_0, \cdots)$ improperly or properly definite, according as the minimum of its $n$-th section $\mathfrak{F}_n$ on the plane $x_0 = 1$, as $n \to \infty$, tends to zero or to a positive quantity. With this notation Theorem 2.18 can be restated in the following form:

*A necessary and sufficient condition that the moment problem* (2.1) *be determined is that at least one of the two quadratic forms*

$$\sum_{i,j=0}^{\infty} \mu_{i+j} x_i x_j, \qquad \sum_{i,j=0}^{\infty} \mu_{i+j+2} x_i x_j$$

*be improperly definite.*

The conditions of Theorems 2.15, 2.16, 2.18 can also be expressed in terms of certain determinants. We start with the quantity $\delta_n(\mu_{-1})$ and observe that

$$W_n(z;\mu_{-1}) \equiv W_n(z) = -\frac{1}{\Delta_n}\begin{vmatrix} 0 & \mu_{-1} & \mu_0 & \mu_1 & \cdots & \mu_{n-1} \\ 1 & \mu_0 & \mu_1 & \mu_2 & \cdots & \mu_n \\ z & \mu_1 & \mu_2 & \mu_3 & \cdots & \mu_{n+1} \\ \cdots & \cdots & \cdots & \cdots & \cdots & \cdots \\ z^n & \mu_n & \mu_{n+1} & \mu_{n+2} & \cdots & \mu_{2n} \end{vmatrix},$$

so that

$$W_n(\mu) = \mu_{-1}.$$

We also have

$$W_n(z) = zV_{n-1}(z) + W_n(0),$$

where

$$V_{n-1}(z) = -\frac{1}{\Delta_n}\begin{vmatrix} 0 & \mu_{-1} & \mu_0 & \cdots & \mu_{n-1} \\ 0 & \mu_0 & \mu_1 & \cdots & \mu_n \\ 1 & \mu_1 & \mu_2 & \cdots & \mu_{n+1} \\ z & \mu_2 & \mu_3 & \cdots & \mu_{n+2} \\ \cdots & \cdots & \cdots & \cdots & \cdots \\ z^{n-1} & \mu_n & \mu_{n+1} & \cdots & \mu_{2n} \end{vmatrix}$$

Thus

$$W_n^2(z) = W_n(z)[zV_{n-1}(z) + W_n(0)] = [Q_{n+1}(z) + 1]V_{n-1}(z) + W_n(z)W_n(0),$$

and, since $Q_{n+1}(z)$ is a quasi-orthogonal polynomial,

$$\begin{aligned}\delta_n = W_n^2(\mu) &= V_{n-1}(\mu) + W_n(\mu)W_n(0) \\ &= V_{n-1}(\mu) + \mu_{-1}W_n(0) = -\frac{1}{\Delta_n}\begin{vmatrix} 0 & \mu_{j-1} \\ \mu_{j-1} & \mu_{i+j}\end{vmatrix}_{i,j=0}^{n}.\end{aligned}$$

To obtain a determinant expression for $\rho_n(z_0)$, where $z_0$ is any given real or complex number, we observe that, by definition,

$$\rho_n(z_0) = \min_{P_n} \overline{P_n}(\mu)P_n(\mu),$$

where the minimum is taken over all polynomials, of degree $\leqq n$, which assume the value 1 at $z = z_0$. Let $S_n(z)$ be the minimizing polynomial. Then we can write

$$P_n(z) = S_n(z) + \lambda p_n(z),$$

where $p_n(z)$ is an arbitrary polynomial, of degree $\leqq n$, vanishing at $z = z_0$, and $\lambda = re^{i\theta}$ is a parameter. The minimum condition can now be written in the form

$$\begin{aligned}S_n(\mu)\overline{S_n(\mu)} &< [S_n(\mu) + \lambda p_n(\mu)][\overline{S_n}(\mu) + \bar{\lambda}\overline{p_n}(\mu)] \\ &= S_n(\mu)\overline{S_n}(\mu) + \bar{\lambda}S_n(\mu)\overline{p_n}(\mu) + \lambda\overline{S_n}(\mu)p_n(\mu) + r^2p_n(\mu)\overline{p_n}(\mu).\end{aligned}$$

In order that this inequality be satisfied for an arbitrary $\lambda \neq 0$ it is clearly necessary and sufficient that

$$S_n(\mu)\overline{p_n}(\mu) = 0,$$

whenever $p_n(z_0) = 0$, or whenever $\overline{p_n}(\overline{z_0}) = 0$.

Since $p_n(z)$ can be written as a linear combination of

$$z - z_0, \quad z^2 - z_0^2, \quad \cdots \quad z^n - z_0^n,$$

the last condition is equivalent to the set of conditions

$$\mu^j S_n(\mu) = \bar{z}_0^j S_n(\mu), \qquad j = 1, 2, \cdots, n,$$

which, together with the condition $S_n(z_0) = 1$, determine $S_n(z)$ uniquely. We easily find that

$$S_n(z) = \begin{vmatrix} 0 & z^j \\ \bar{z}_0^i & \mu_{i+j} \end{vmatrix}_{i,j=0}^{n} \Bigg/ \begin{vmatrix} 0 & z_0^j \\ \bar{z}_0^i & \mu_{i+j} \end{vmatrix}_{i,j=0}^{n}.$$

Since

$$S_n(\mu)\overline{S_n}(\mu) = S_n(z_0)S_n(\mu) = S_n(\mu),$$

we finally have

$$\rho_n(z_0) = S_n(\mu)\overline{S_n}(\mu) = -\Delta_n \Bigg/ \begin{vmatrix} 0 & z_0^j \\ \bar{z}_0^i & \mu_{i+j} \end{vmatrix}_{i,j=0}^{n}.$$

In particular,

$$\rho_n(0) = \begin{vmatrix} \mu_0 & \mu_1 & \cdots & \mu_n \\ \cdots & \cdots & \cdots & \cdots \\ \mu_n & \mu_{n+1} & \cdots & \mu_{2n} \end{vmatrix} \Bigg/ \begin{vmatrix} \mu_2 & \cdots & \mu_{n+1} \\ \cdots & \cdots & \cdots \\ \mu_{n+1} & \cdots & \mu_{2n} \end{vmatrix}.$$

In the same way we find

$$\rho_{n-1}^{(2)}(0) = \begin{vmatrix} \mu_2 & \cdots & \mu_n \\ \cdots & \cdots & \cdots \\ \mu_n & \cdots & \mu_{2n} \end{vmatrix} \Bigg/ \begin{vmatrix} \mu_4 & \cdots & \mu_{n+2} \\ \cdots & \cdots & \cdots \\ \mu_{n+2} & \cdots & \mu_{2n} \end{vmatrix}.$$

Substituting this in Theorem 2.18 and taking the product $\rho_n(0)\rho_{n-1}^{(2)}(0)$, we obtain finally

THEOREM 2.19. *A necessary and sufficient condition that the moment problem* (2.1) *be determined is that the monotonically decreasing ratio of two determinants*

$$\begin{vmatrix} \mu_0 & \cdots & \mu_n \\ \cdots & \cdots & \cdots \\ \mu_n & \cdots & \mu_{2n} \end{vmatrix} \Bigg/ \begin{vmatrix} \mu_4 & \cdots & \mu_{n+2} \\ \cdots & \cdots & \cdots \\ \mu_{n+2} & \cdots & \mu_{2n} \end{vmatrix}$$

*tend to zero, as* $n \to \infty$.

**24.** Without going into an exposition of the classical memoir of Stieltjes, [5], we shall mention here very briefly some distinctions between the theories of the Stieltjes and of the Hamburger moment problems.

We assume here that the Stieltjes moment problem corresponding to the sequence of moments $\{\mu_n\}$, $n = 0, 1, \cdots$, has a solution whose spectrum is not a finite set, so that not only

$$\Delta_n = |\mu_{i+j}|_{i,j=0}^{n} > 0, \qquad n = 0, 1, 2, \cdots,$$

but also

$$\Delta_n^{(1)} = |\mu_{i+j+1}|_{i,j=0}^{n} > 0, \qquad n = 0, 1, 2, \cdots.$$

Then using some well known facts from the theory of continued fractions [Perron, 1] it is readily shown that the associated continued fraction

$$\frac{\beta_0|}{|z-\alpha_1} - \frac{\beta_1|}{|z-\alpha_2} - \frac{\beta_2|}{|z-\alpha_3} - \cdots$$

can be transformed into the corresponding continued fraction

$$\frac{1|}{|l_1 z} + \frac{1|}{|l_2} + \frac{1|}{|l_3 z} + \frac{1|}{|l_4} + \cdots,$$

where

$$\text{(2.67)} \qquad \begin{aligned} l_{2n} &= -\frac{\Delta_{n-1}^2}{\Delta_{n-1}^{(1)}\Delta_{n-2}^{(1)}} < 0, \qquad n = 2, 3, \cdots; \qquad l_2 = -\frac{\mu_0^2}{\mu_1}; \\ l_{2n+1} &= \frac{\Delta_{n-1}^{(1)2}}{\Delta_{n-1}\Delta_n} > 0, \qquad n = 1, 2, \cdots; \qquad l_1 = \frac{1}{\mu_0}. \end{aligned}$$

This continued fraction is such that the fractions $\dfrac{\chi_n(z)}{\omega_n(z)}$ are its approximants of even order, while the fractions $\dfrac{C_{n+1}(z)}{D_{n+1}(z)}$ are approximants of odd order. A simple way of proving this is to use the quasi-orthogonality properties of the polynomials $\omega_n(z)$ and $D_{n+1}(z)$.

Write

$$\omega_n(z) \equiv \omega'_{2n}(z), \qquad \frac{\chi_n(z)}{\omega_n(z)} \equiv \frac{\chi'_{2n}(z)}{\omega'_{2n}(z)}, \qquad n = 0, 1, 2, \cdots,$$

$$D_{n+1}(z) \equiv \omega'_{2n+1}(z), \qquad \frac{C_{n+1}(z)}{D_{n+1}(z)} \equiv \frac{\chi'_{2n+1}(z)}{\omega'_{2n+1}(z)}, \qquad n = 0, 1, 2, \cdots.$$

Again, by using simple properties of continued fractions, Stieltjes proves* that for a negative $z$ the sequence $\left\{\dfrac{\chi'_{2n}(z)}{\omega'_{2n}(z)}\right\}$ is decreasing, the sequence $\left\{\dfrac{\chi'_{2n+1}(z)}{\omega'_{2n+1}(z)}\right\}$ is

* We do not reproduce Stieltjes' proof here since we shall give another proof, based on a different principle, in Chapter IV. It should be observed that Stieltjes considers expressions of the type $\displaystyle\int_0^\infty \frac{d\psi(t)}{z+t}$, so that his functions are related to those discussed here by the general transformation formula $-f(-z)$.

increasing, and each term of the second sequence is less than each term of the first. From this and from the fact that each of these sequences is uniformly bounded in every domain of the complex plane which is at a positive distance from the segment $[0, \infty)$, by an application of the Stieltjes-Vitali theorem, it is easy to show that each of the above two sequences converges separately on the whole plane cut along the segment $[0, \infty)$ On the other hand, let $\psi(t)$ be any solution of the Stieltjes moment problem. We have two identities [Perron, 1]:

$$\omega'_{2n}(z)^2 \int_0^\infty \frac{d\psi(t)}{z-t} - \omega'_{2n}(z) \int_0^\infty \frac{\omega'_{2n}(z) - \omega'_{2n}(t)}{z-t}\, d\psi(t)$$

$$- \int_0^\infty \omega'_{2n}(t) \frac{\omega'_{2n}(z) - \omega'_{2n}(t)}{z-t}\, d\psi(t) = \int_0^\infty \frac{\omega'_{2n}(t)^2}{z-t}\, d\psi(t),$$

$$\omega'_{2n+1}(z)^2 \int_0^\infty \frac{d\psi(t)}{z-t} - \omega'_{2n+1}(z) \int_0^\infty \frac{\omega'_{2n+1}(z) - \omega'_{2n+1}(t)}{z-t}\, d\psi(t)$$

$$- z \int_0^\infty \omega'_{2n+1}(t) \frac{\dfrac{\omega'_{2n+1}(z)}{z} - \dfrac{\omega'_{2n+1}(t)}{t}}{z-t}\, d\psi(t)$$

$$= z \int_0^\infty \frac{\omega'_{2n+1}(t)^2}{t} \frac{d\psi(t)}{z-t}.$$

Let $z = z_0$ be any negative number. Then, using the quasi-orthogonal properties of the polynomials $\omega'_{2n}(z)$, $\omega'_{2n+1}(z)$ and the integral representation of the corresponding numerators we immediately conclude that [Perron, 1]

$$\int_0^\infty \frac{d\psi(t)}{z_0 - t} - \frac{\chi'_{2n}(z_0)}{\omega'_{2n}(z_0)} = \frac{1}{\omega'_{2n}(z_0)^2} \int_0^\infty \frac{\omega'_{2n}(t)^2}{z_0 - t}\, d\psi(t) < 0,$$

$$\frac{\chi'_{2n+1}(z_0)}{\omega'_{2n+1}(z_0)} - \int_0^\infty \frac{d\psi(t)}{z_0 - t} = -\frac{z_0}{\omega'_{2n+1}(z_0)^2} \int_0^\infty \frac{\omega'_{2n+1}(t)^2}{t} \frac{d\psi(t)}{z_0 - t} < 0.$$

Thus we see that the Stieltjes moment problem corresponding to the sequence of moments $\{\mu_n\}$, $n = 0, 1, 2, \cdots$, is or is not determined according as the sequences $\left\{\dfrac{\chi'_{2n}(z)}{\omega'_{2n}(z)}\right\}$, $\left\{\dfrac{\chi'_{2n+1}(z)}{\omega'_{2n+1}(z)}\right\}$ converge to the same limit or not, for all $z$ which are not in the interval $[0, \infty)$, or, which is the same, according as the corresponding continued fraction converges or does not converge for these values of $z$. This continued fraction can be written [Perron, 1] as

$$-\frac{1\,|}{|\,-l_1 z} + \frac{1\,|}{|\,-l_2} + \frac{1\,|}{|\,-l_3 z} + \frac{1\,|}{|\,-l_4} + \cdots,$$

and if $z$ is negative, a well known [Perron, 1] necessary and sufficient condition for the convergence of such a fraction is simply that the series $\sum_{n=1}^{\infty} | l_n | = \infty$. We thus arrive at the following fundamental theorem of Stieltjes.

THEOREM 2.20. *The Stieltjes moment problem corresponding to a given sequence of moments* $\{\mu_n\}$, $n = 0, 1, 2, \cdots$, *(for which it is assumed to have a solution) is determined or not according as the series* $\sum_{n=1}^{\infty} | l_n |$ *is divergent or convergent, where the* $l_n$ *are given by* (2.67). *It is assumed here that* $\Delta_n > 0$, $\Delta_n^{(1)} > 0$, $n = 0, 1, 2, \cdots$.

By an argument quite similar to that used in **9** Stieltjes proves that

$$(2.68)\qquad \begin{aligned} \left| I(x;\psi) - \frac{\chi'_{2n}(x)}{\omega'_{2n}(x)} \right| &= \min_{P_n} \int_0^{\infty} P_n(t)^2 \frac{d\psi(t)}{t-x}, \\ |x| \left| I(x;\psi) - \frac{\chi'_{2n+1}(x)}{\omega'_{2n+1}(x)} \right| &= \min_{P_n} \int_0^{\infty} P_n(t)^2 \frac{t\, d\psi(t)}{t-x}, \end{aligned}$$

where $x < 0$ and the minimum is taken over all polynomials $P_n(t)$, of degree $\leqq n$, assuming the value 1 at $t = x$.

As was pointed out above, in the Stieltjes case, if $x < 0$, both sequences $\left\{\frac{\chi'_{2n}(x)}{\omega'_{2n}(x)}\right\}$, $\left\{\frac{\chi'_{2n+1}(x)}{\omega'_{2n+1}(x)}\right\}$, of the odd and even approximants of the corresponding continued fraction, converge. On the other hand, by (2.68), a necessary and sufficient condition that $\frac{\chi'_{2n}(x)}{\omega'_{2n}(x)} \to I(x;\psi)$ is that the distribution $\frac{d\psi(t)}{t-x}$ correspond to a determined Hamburger moment problem. This, combined with a sufficient condition for the validity of the Parseval formula in the Stieltjes case (p. 66) and with Theorem 2.14, shows that *the distribution function corresponding to* $\omega_n(z) \equiv \omega'_{2n}(z)$ *tends, as* $n \to \infty$, *to an extremal solution of the moment problem* (2.1). *The same holds for the sequence of distribution functions corresponding to the quasi-orthogonal polynomial* $\omega'_{2n+1}(z)$, as follows from Corollary 2.5, if we take there $\lambda = 0$. As a consequence of this, *if we denote by* $\psi(t)$ *any one of these two solutions, the Parseval formula holds for every function* $f \in L_\psi^2$.

**25.** The following result [Hamburger, 3] should be compared with the result on p. 70.

*A necessary and sufficient condition that the Stieltjes moment problem corresponding to the sequence of moments* $\{\mu_n\}$, $n = 0, 1, 2, \cdots$, *be determined is that at least one of the definite forms*

$$\sum_{i,j=0}^{\infty} \mu_{i+j}\, x_i\, x_j\,, \qquad \sum_{i,j=0}^{\infty} \mu_{i+j+1}\, x_i\, x_j$$

*be improperly definite.*

Thus, in the case of the Stieltjes moment problem the simple convergence of the corresponding continued fraction plays the same role as the complete convergence of the associated continued fraction in the case of the Hamburger moment problem. It may well happen that a given sequence of moments $\{\mu_n\}$, $n = 0, 1, 2, \cdots$, corresponds to a determined Stieltjes moment problem, and to an indeterminate Hamburger moment problem.* We refer to Stieltjes, [5] for numerous examples of developments in continued fractions, and to Stieltjes, [5] and Hamburger [3] for the discussion of various cases of convergence of continued fractions which are not completely convergent, or which present various other interesting convergence phenomena.

Another interesting distinction between the theories of the Stieltjes and the Hamburger moment problems is in the evaluation of the largest possible concentration of mass at a given point $x_0$. We know that in the case of the Hamburger moment problem the largest mass which can be concentrated at a point $x_0$ is $\rho(x_0)$. Stieltjes proves that if his moment problem is indeterminate, this estimate should be compared with the estimate $\frac{1}{x_0}\rho^{(1)}(x_0)$, where $\rho^{(1)}(x_0)$ is determined by the sequence of moments $\{\mu_n\}$, $n = 1, 2, \cdots$, in the same way as $\rho(x_0)$ is determined by the sequence $\{\mu_n\}$, $n = 0, 1, 2, \cdots$. In fact, Stieltjes proves that the estimate $\rho(x_0)$ holds when $x_0$ is in one of the intervals $[0, \lambda_1]$, $(\theta_1, \lambda_2)$, $(\theta_2, \lambda_3)$, $\cdots$, while the estimate $\frac{1}{x_0}\rho_0^{(1)}(x_0)$ holds when $x_0$ is in one of the intervals $(\lambda_1, \theta_1)$, $(\lambda_2, \theta_2)$, $(\lambda_3, \theta_3)$, $\cdots$; these estimates coincide when $x_0 = \lambda_1$, $\theta_1, \lambda_2, \theta_2, \cdots$. Here $0 < \theta_1 < \theta_2 < \cdots$ are the roots of the entire function

$$q_1(z) = \lim_n D_{n+1}(z),$$

while $\lambda_1 < \lambda_2 < \lambda_3 < \cdots$ are the roots of the entire function

$$q(z) = \lim_n \frac{\omega_n(z)}{\omega_n(0)}.$$

* Such is the case of the continued fraction

$$-\frac{1|}{|-\epsilon z} + \frac{1|}{|1} + \frac{1|}{|-\epsilon^2 z} + \frac{1|}{|1/2} + \frac{1|}{|-\epsilon^3 z} + \frac{1|}{|1/3} + \cdots$$

which was communicated orally to the authors by J. V. Uspensky.

## CHAPTER III

# VARIOUS MODIFICATIONS OF THE MOMENT PROBLEM.

**1.** Part of this chapter is devoted to the exposition of some problems stated by Tchebycheff and solved later by A. Markoff. The first of these problems may be stated as follows:

PROBLEM (M). *Given* $(n + 1)$ *constants* $\mu_0, \mu_1, \cdots, \mu_n$, *for which the reduced moment problem*

$$\int_{-1}^{1} t^{\nu} \, d\psi(t) = \mu_{\nu}, \qquad \nu = 0, 1, 2, \cdots, n, \tag{3.1}$$

*admits of a solution* $\psi(t)$. *Let* $x$ *be a given point of the interval* $(-1, 1)$, *and* $f(t)$ *a given function. Determine*

$$\inf_{\psi} \int_{-1}^{x} f(t) \, d\psi(t), \qquad \sup_{\psi} \int_{-1}^{x} f(t) \, d\psi(t), \tag{3.2}$$

*when* $\psi(t)$ *ranges over all solutions of the reduced moment problem* (3.1).

We give here a brief exposition of a solution of this problem which is due to Possé [1] who somewhat simplified the original proof of Markoff [1]. In a later paper Markoff extends his results to the case of an infinite interval [15] (Cf. also Achyeser and Krein, [6]). It is clear that the case of a general finite interval $(a, b)$ can be reduced to that treated here by a simple linear transformation.

We first derive necessary and sufficient conditions for the existence of solutions of the moment problem (3.1), in a form different from that used in Chapter I.

THEOREM 3.1. *A necessary and sufficient condition that the moment problem* (3.1) *have a solution is that, in case* $n = 2m$, *both quadratic forms*

$$\sum_{i,j=0}^{m} \mu_{i+j} x_i x_j, \qquad \sum_{i,j=0}^{m} (\mu_{i+j} - \mu_{i+j+2}) x_i x_j \tag{3.3}$$

*be non-negative, while in case* $n = 2m - 1$, *both quadratic forms*

$$\sum_{i,j=0}^{m} (\mu_{i+j} + \mu_{i+j+1}) x_i x_j, \qquad \sum_{i,j=0}^{m-1} (\mu_{i+j} - \mu_{i+j+1}) x_i x_j \tag{3.4}$$

*be non-negative.*

The proof is based on the fact [Pólya-Szegö 1, vol. II, p. 82] that a polynomial which is non-negative on the interval $[-1, 1]$ admits of a representation

$$\left(\sum_{i=0}^{m} x_i t^i\right)^2 + (1 - t^2)\left(\sum_{i=0}^{m-1} y_i t^i\right)^2,$$

if it is of degree $2m$, and of a representation

$$(1 + t)\left(\sum_{i=0}^{m-1} x_i t^i\right)^2 + (1 - t)\left(\sum_{i=0}^{m-1} y_i t^i\right)^2,$$

if it is of degree $2m - 1$. Since this proof goes along the same (and even simpler) lines as in Chapter I, it may be omitted.

**2.** *The sequence* $(\mu_0, \cdots, \mu_n)$ *is called positive (non-negative) if the fact that* $P_n(t)$ *is a polynomial, of degree* $\leqq n$, *which is not identically zero and is* $\geqq 0$ *on* $[-1, 1]$, *implies that* $\mu(P_n) > 0 (\mu(P_n) \geqq 0)$. [Hausdorff, 1]. It is clear that in order that the sequence $(\mu_0, \cdots, \mu_n)$ be positive (non-negative) it is necessary and sufficient that the corresponding quadratic forms be also positive (non-negative). The condition of positiveness will be assumed in the following discussion unless explicitly stated to the contrary. Otherwise the solution $\psi(t)$ of (3.1) is uniquely determined and the case does not present any interest.

Let $\psi(t)$ be any solution of the moment problem (3.1). We shall have to discuss the ortho-normal polynomials

$$\omega_n(z), \qquad \omega_n^{(-1)}(z), \qquad \omega_n^{(+1)}(z), \qquad \omega_{n-1}^{(-1,1)}(z),$$

determined respectively by the continued-fraction expansions of the integrals

$$\int_{-1}^{1} \frac{d\psi(t)}{z - t}, \qquad \int_{-1}^{1} \frac{(1 + t)\, d\psi(t)}{z - t}, \qquad \int_{-1}^{1} \frac{(1 - t)\, d\psi(t)}{z - t}, \qquad \int_{-1}^{1} \frac{(1 - t^2)\, d\psi(t)}{z - t}.$$

We denote the corresponding numerators, respectively, by

$$\chi_n(z), \qquad \chi_n^{(-1)}(z), \qquad \chi_n^{(+1)}(z), \qquad \chi_{n-1}^{(-1,1)}(z).$$

As we know, the polynomial $\omega_n(z)$ is completely determined by the moments $\mu_0, \mu_1, \cdots, \mu_{2n-1}$. Observe, further, that the $\nu$-th moment of the distributions $(1 + t)\, d\psi(t)$, $(1 - t)\, d\psi(t)$, $(1 - t^2)\, d\psi(t)$ is, respectively, $\mu_\nu + \mu_{\nu+1}$, $\mu_\nu - \mu_{\nu+1}$, $\mu_\nu - \mu_{\nu+2}$. It readily follows that $\omega_n^{(\pm 1)}(z)$ is completely determined by the moments $\mu_0, \mu_1, \cdots, \mu_{2n}$, and $\omega_{n-1}^{(-1,1)}(z)$ by $\mu_0, \mu_1, \cdots, \mu_{2n-1}$.

In the preceding chapter we have proved the reality and simplicity of roots of quasi-orthogonal polynomials. Here, using the same method as in the proof of Lemma 2.8, it can be proved, in addition, that the roots of orthogonal polynomials are always in the interval $(-1, 1)$, the same being also true for quasi-orthogonal polynomials with the possible exception of at most one root. Thus, the roots of $\omega_n(z)$, $\omega_n^{(-1)}(z)$, $\omega_n^{(+1)}(z)$, $\omega_{n-1}^{(-1,1)}(z)$ are all in $(-1, 1)$. By analogous methods it can be shown that the roots of $\omega_n^{(-1)}(z)$ are separated by those of $(z - 1)\omega_n^{(+1)}(z)$, the roots of $\omega_n^{(+1)}(z)$ are separated by those of $(z + 1)\omega_n^{(-1)}(z)$, the roots of $\omega_n(z)$ are separated by those of $(z^2 - 1)\omega_{n-1}^{(-1,1)}(z)$ and that the roots of $\omega_{n+1}^{(-1,1)}(z)$ are separated by those of $\omega_n(z)$.

**3.** An important step in solving problem $(M)$ is the proof of the existence of certain special solutions of the moment problem (3.1) which are step-functions such that the spectrum of each contains the given point $x$.

LEMMA 3.1. *Let $x$ be a given point in $(-1, 1)$ distinct from the roots of $\omega_n(z)$, $\omega_n^{(-1)}(z)$, $\omega_n^{(+1)}(z)$, $\omega_{n-1}^{(-1,1)}(z)$. Consider the moment problem* (3.1), *and let $q(z)$ denote one of the following four polynomials*

$$\text{(i)} \qquad \begin{vmatrix} (z+1)\omega_m^{(-1)}(z), & (z-1)\omega_m^{(+1)}(z) \\ (x+1)\omega_m^{(-1)}(x), & (x-1)\omega_m^{(+1)}(x) \end{vmatrix},$$

*if $n = 2m$ and* $\operatorname{sgn} \omega_m^{(-1)}(x) = \operatorname{sgn} \omega_m^{(+1)}(x)$,

$$\text{(ii)} \qquad (z^2-1) \begin{vmatrix} \omega_m^{(-1)}(z), & \omega_m^{(+1)}(z) \\ \omega_m^{(-1)}(x), & \omega_m^{(+1)}(x) \end{vmatrix},$$

*if $n = 2m$ and* $\operatorname{sgn} \omega_m^{(-1)}(x) = -\operatorname{sgn} \omega_m^{(+1)}(x)$

$$\text{(iii)} \qquad (z+1) \begin{vmatrix} \omega_m(z), & (z-1)\omega_{m-1}^{(-1,1)}(z) \\ \omega_m(x), & (x-1)\omega_{m-1}^{(-1,1)}(x) \end{vmatrix},$$

*if $n = 2m - 1$ and* $\operatorname{sgn} \omega_m(x) = \operatorname{sgn} \omega_{m-1}^{(-1,1)}(x)$,

$$\text{(iv)} \qquad (z-1) \begin{vmatrix} \omega_m(z), & (z+1)\omega_{m-1}^{(-1,1)}(z) \\ \omega_m(x), & (x+1)\omega_{m-1}^{(-1,1)}(x) \end{vmatrix},$$

*if $n = 2m - 1$ and* $\operatorname{sgn} \omega_m(x) = -\operatorname{sgn} \omega_{m-1}^{(-1,1)}(x)$.

*Let $p(z)$ be the numerator corresponding to $q(z)$, $\{x_j\}$ the sequence of roots of $q(z)$, and*

$$H_j = \frac{p(x_j)}{q'(x_j)}.$$

*The sequence $\{x_j\}$ always includes $x$, and may include also one or both end-points $\pm 1$.*

*We always have $H_j > 0$. The step-function whose spectrum coincides with $\{x_j\}$ and which has the mass $H_j$ at $x_j$ is a solution of the moment problem* (3.1). *Finally, if $\psi(t)$ is any solution of* (3.1), *the approximate quadrature formula*

$$\int_{-1}^{1} P_n(t)\, d\psi(t) = \sum_j H_j P_n(x_j) \tag{3.5}$$

*holds for an arbitrary polynomial $P_n(t)$, of degree $\leqq n$.*

We sketch a proof of Lemma 3.1 only in the case (i); other cases can be treated by similar methods, and we refer for proofs to Possé [1]. By Lemma 2.7, we can always find a quasi-orthogonal polynomial $q(z)$, of degree $(m + 1)$, which has a mass $\dfrac{p(x)}{q(x)} = \rho_m(x) = H$ concentrated at $x$, and masses $\dfrac{p(x_j)}{q(x_j)} = \rho_m(x_j) = H_j$ concentrated at other roots $x_1, \cdots, x_m$ of $q(z)$, and such that formula (3.5) holds for any polynomial $P_{2m}(z)$ of degree $\leqq 2m$. It remains to investigate under what conditions *all* the roots $x_1, \cdots, x_m$ will be in $[-1, 1]$. From the preceding we know only that all these roots, except possibly one, satisfy this condition. Now, since the polynomials $(z+1)\omega_m^{(-1)}(z)$, $(z-1)\omega_m^{(+1)}(z)$ are quasi-orthogonal polynomials of degree $(m+1)$, we may write, by Lemma 2.6,

$$(3.6) \qquad q(z) = \begin{vmatrix} (z+1)\omega_m^{(-1)}(z), & (z-1)\omega_m^{(+1)}(z) \\ (x+1)\omega_m^{(-1)}(x), & (x-1)\omega_m^{(+1)}(x) \end{vmatrix},$$

and a necessary and sufficient condition that all roots of $q(z)$ should be in $(-1, 1)$ is, clearly,

$$(3.7) \qquad \operatorname{sgn} (-1)^{m+1} q(-1) = \operatorname{sgn} q(1).$$

In view of the obvious inequalities

$$(x-1)\omega_m^{(-1)}(1) < 0, \qquad (-1)^{m+1}(x+1)\omega_m^{(+1)}(-1) < 0,$$

formula (3.6) shows that condition (3.7) is equivalent to the condition

$$\operatorname{sgn} \omega_m^{(-1)}(x) = \operatorname{sgn} \omega_m^{(+1)}(x),$$

which is precisely our condition in the case (i).

We also need the following lemma due to A. Markoff [1] (See also Possé [1]).

LEMMA 3.2. *Let $f(t)$ be continuous in $[a, b]$, together with its $(m+1)$ first derivatives and let*

*$f(t) > 0, f^{(k)}(t) \geqq 0$, $k = 1, 2, \cdots, (m+1)$, in $[a, b]$. Let $P_m(t)$ be a polynomial, of degree $\leqq m$, and $c$ a given point in $(a, b)$. Let $m_1$ be the number of roots in $(a, c)$ of the equation $f(t) = P_m(t)$ and $m_2$ the number of roots in $(c, b)$ of the equation $P_m(t) = 0$. Then $m_1 + m_2 \leqq m + 1$. Here multiple roots are counted according to their multiplicity.*

This Lemma is readily proved by a repeated application of Rolle's theorem.

**4.** We are now in a position to give a solution of problem (M), at least when the given function $f(t)$ is subjected to certain restrictions.

THEOREM 3.2. *Let $f(t)$ be a given function continuous on $[-1, 1]$ together with its first $(n+1)$ derivatives, and let*

$$(3.8) \quad f(t) > 0, \qquad f^{(k)}(t) \geqq 0, \qquad k = 1, 2, \cdots, (n+1), \quad \text{in } [-1, 1].$$

*Let $\psi(t)$ be any solution of the moment problem* (3.1). *Then*

$$(3.9) \qquad \sum_{x_i < x} H_i f(x_i) \leqq \int_{-1}^{x} f(t)\, d\psi \leqq \sum_{x_i \leqq x} H_i f(x_i).$$

Again we shall give a proof only in the case (i) of Lemma 3.1 and refer to Possé [1] for analogous proofs in the remaining cases (ii, iii, iv). Assume that conditions of the case (i) are satisfied and that $f(t)$ is a given function continuous together with the derivatives $f^{(k)}(t)$, $k = 1, 2, \cdots, 2m+1$, in $[-1, 1]$, and such that

$$f(t) > 0, \qquad f^{(k)}(t) \geqq 0, \qquad k = 1, 2, \cdots, (2m+1), \quad \text{in } [-1, 1].$$

Denote by

$$x_1 < x_2 < \cdots < x_k < x < x_{k+1} < \cdots < x_m, \qquad -1 < x_1 < x_m < 1$$

all roots of the equation $q(z) = 0$. The cases where $x < x_1$ or $x > x_m$ are treated in a similar manner.

Construct two polynomials $\Phi_1(z)$, $\Phi_2(z)$, each of degree $\leqq 2m$, such that

$$\Phi_1(t) \leqq f(t) \leqq \Phi_2(t) \quad \text{in } [-1, x],$$

$$\Phi_1(t) \leqq 0 \text{ in } [x, 1], \qquad \Phi_2(t) \geqq 0 \quad \text{in } [x, 1],$$

and

$$\Phi_1(t) = f(t), \qquad \text{for } t = x_1, x_2, \cdots, x_k,$$

$$\Phi_2(t) = f(t), \qquad \text{for } t = x_1, x_2, \cdots, x_k, x,$$

$$\Phi_1(t) = \Phi_2(t) = 0, \qquad \text{for } t = x_{k+1}, \cdots, x_m.$$

Assuming, for the moment, the possibility of this construction, we have

$$\int_{-1}^{1} \Phi_1(t)\, d\psi \leqq \int_{-1}^{1} f(t)\, d\psi \leqq \int_{-1}^{1} \Phi_2(t)\, d\psi,$$

from which, by (3.5), (3.9) follows, so that Theorem 3.2 is proved.

Now let

$$\Phi_0^{(1)}(z) = \sum_{x_j < x} \frac{q(z)}{(z - x_j)q'(x_j)} f(x_j), \qquad \Phi_0^{(2)}(z) = \sum_{x_j \leqq x} \frac{q(z)}{(z - x_j)q'(x_j)} f(x_j),$$

$$\Phi_1(z) = \Phi_0^{(1)}(z) + q(z)P_{m-1}^{(1)}(z), \qquad \Phi_2(z) = \Phi_0^{(2)}(z) + q(z)P_{m-1}^{(2)}(z),$$

where $P_{m-1}^{(1)}(z)$, $P_{m-1}^{(2)}(z)$ are polynomials, of degree $\leqq (m - 1)$, determined, respectively, by the conditions

$$\begin{cases} \Phi_i'(x_j) = f'(x_j), & x_j < x, \\ \Phi_i'(x_j) = 0, & x_j > x, \end{cases} \qquad i = 1, 2.$$

Consider, to be definite, the polynomial $\Phi_2(z)$. The equation $f(z) = \Phi_2(z)$ has roots $x_1, x_2, \cdots, x_k$, of multiplicity $\geqq 2$, and the root $x$ of multiplicity $\geqq 1$, while the equation $\Phi_2(z) = 0$ has roots $x_{k+1}, \cdots, x_m$, of multiplicity $\geqq 2$. Since $2k + 1 + 2(m - k) = 2m + 1$, we see, by Lemma 3.2, that $x_1, \cdots, x_k$ are double roots of the equation $f(z) = \Phi_2(z)$ in $(0, x)$, $x_{k+1}, \cdots, x_m$ are double roots of the equation $\Phi_2(z) = 0$ in $(x, 1)$, $x$ is a simple root of the equation $f(z) = \Phi_2(z)$, and that the equations in question have no more roots in the corresponding intervals. Now $\Phi_2(x_{k+1}) - f(x_{k+1}) = -f(x_{k+1}) < 0$, so that $\Phi_2(t) - f(t) < 0$ in $(x, x_{k+1})$. Since $z = x$ is a simple root and $x_1, \cdots, x_k$ are double roots of the equation $\Phi_2(z) - f(z) = 0$, we see that $\Phi_2(t) - f(t) \geqq 0$ in the whole interval $[-1, x]$. On the other hand, $\Phi_2(z) > 0$ in $(x, x_{k+1})$, and since $x_{k+1}, \cdots, x_m$ are double roots of the equation $\Phi_2(z) = 0$, it follows that $\Phi_2(z) \geqq 0$ in the whole interval $[x, 1]$. Thus, the polynomial $\Phi_2(z)$ has all desired properties, and the same can be proved in a similar manner for the polynomial $\Phi_1(z)$. This completes the proof of Theorem 3.2.

Analogous results hold in the case where $x$ is a root of one of the polynomials $\omega_m(z)$, $\omega^{(-1)}(z)$, $\omega^{(+1)}(z)$, $\omega_{m-1}^{(-1,1)}(z)$ or $x = \pm 1$ [Possé, 1].

By specifying the function $f(t)$ and the point $x$ in the above discussion, we may obtain various inequalities useful in applications to probability and statistics. An interesting case is $f(t) = t^{n+1}$, $x = 1$, which yields bounds for the moment $\mu_{n+1}$ of a distribution function in terms of the preassigned moments $\mu_0, \mu_1, \cdots, \mu_n$. These bounds are best derived by means of Theorem 3.2. Thus, for instance, in case $(a, b) \equiv (-1, 1)$,

$$\frac{\mu_2(\mu_0 - \mu_2) - \mu_1(\mu_1 - \mu_2)}{\mu_0 - \mu_1} \geqq \mu_3 \geqq \frac{\mu_2(\mu_2 - \mu_0) + \mu_1(\mu_1 + \mu_2)}{\mu_0 + \mu_1};$$

Recently, A. Wald [1] has generalized the problem concerning bounds for the moments and also Tchebycheff inequalities. (Cf. also M. Fréchet, [1]).

**5.** Another class of problems in which Markoff was much interested deals with conditions for the existence of an absolutely continuous distribution $d\psi(t) = \varphi(t)dt$ of a moment problem (for a finite or infinite interval, and also for the trigonometric moment problem), where $\varphi(t)$, in addition to integrability, is subject to various additional restrictions [Markoff, 10, 11, 13, 14]. The results of Markoff have been modernized and extended in several papers by Achyeser and Krein [2, 4] and also in their recent book [6] to which we have already referred (see also papers by Verblunsky [2, 3, 4, 5], which apparently were written without the knowledge of the previous work of Markoff and of Achyeser and Krein). We give here a brief exposition of one typical problem following the treatment by Achyeser and Krein, and referring for a systematic treatment of the whole subject to their book [6].

THEOREM 3.3. *A necessary and sufficient condition that the reduced moment problem*

$$\mu_\nu = \int_{-\infty}^{\infty} t^\nu \varphi(t)\, dt, \qquad \nu = 0, 1, \cdots, 2n, \qquad \mu_0 > 0, \tag{3.10}$$

*admit of an integrable solution $\varphi(t)$ satisfying almost everywhere the condition*

$$0 \leqq \varphi(t) < L \tag{3.11}$$

*is that the sequence of moments*

$$\mu_0(L), \cdots, \mu_{2n}(L)$$

*defined by the expansion*

$$\exp\left[\frac{1}{L}\left(\frac{\mu_0}{z} + \cdots + \frac{\mu_{2n}}{z^{2n+1}}\right)\right] \equiv 1 + \frac{\mu_0(L)}{z} + \cdots + \frac{\mu_{2n}(L)}{z^{2n+1}} + \cdots \tag{3.12}$$

*be positive over* $(-1, 1)$, *or, which is the same, that the quadratic form*

$$\sum_{i,j=0}^{n} \mu_{i+j}(L)x_i x_j \tag{3.13}$$

*be positive.*

First, we prove the sufficiency of the above condition. If the quadratic form (3.13) is positive, then for every $N < L$ and sufficiently near to $L$, the quadratic form

$$\sum_{i,j=0}^{n} \mu_{i+j}(N)x_i x_j$$

is also positive. Fix the value of $N$ and write

$$\mu_\nu(N) = \mu'_\nu, \qquad \nu = 0, 1, 2, \cdots, 2n.$$

We have seen in II.**8** that we can always find a quasi-orthogonal polynomial $q(z)$, of degree $(n + 1)$, and the corresponding distribution function $\psi_n(t)$ of order $n$, determined by the sequence $\{\mu'_\nu\}$ such that $\psi_n(t)$ is a step-function which has points of increase at the roots $x_0, x_1, x_2 \cdots, x_n$ of $q(z)$ with the concentrated masses $\rho_n(x_0), \cdots, \rho_n(x_n)$ respectively. We have, then,

$$\sum_{j=0}^{n} \rho_n(x_j)x_j^\nu = \mu'_\nu, \qquad \nu = 0, 1, \cdots, 2n,$$

$$1 + \sum_{j=0}^{n} \frac{\rho_n(x_j)}{z - x_j} \equiv \frac{s(z)}{q(z)} = 1 + \frac{\mu'_0}{z} + \cdots + \frac{\mu'_{2n}}{z^{2n+1}} + \cdots.$$

If we write

$$q(z) = (z - x_0) \cdots (z - x_n), \qquad s(z) = (z - y_0) \cdots (z - y_n),$$

it is readily seen that

$$y_0 < x_0 < y_1 < x_1 < \cdots < y_n < x_n .$$

Now introduce the step function assuming, except for removable discontinuities, only values 0 and $N$,

$$\varphi_N(t) \equiv \frac{N}{2}\left\{1 - \operatorname{sgn} \frac{s(t)}{q(t)}\right\}.$$

For a sufficiently large positive $z$ we have

$$\int_{-\infty}^{\infty} \frac{\varphi_N(t)\, dt}{z - t} = N \log \frac{s(z)}{q(z)}.$$

On the other hand, on replacing $L$ by $N$ in (3.12) and taking logarithms,

$$\frac{\mu_0}{z} + \frac{\mu_1}{z^2} + \cdots + \frac{\mu_{2n}}{z^{2n}} = N \log\left[1 + \frac{\mu'_0}{z} + \frac{\mu'_1}{z^2} + \cdots + \frac{\mu'_{2n}}{z^{2n+1}} + \cdots\right],$$

which, compared with the preceding results, shows that

$$N \log \frac{s(z)}{q(z)} = \int_{-\infty}^{\infty} \frac{\varphi_N(t)\,dt}{z-t} = \frac{\mu_0}{z} + \frac{\mu_1}{z^2} + \cdots + \frac{\mu_{2n}}{z^{2n+1}} + \cdots.$$

Thus

$$\mu_\nu = \int_{-\infty}^{\infty} t^\nu \varphi_N(t)\,dt, \qquad \nu = 0, 1, \cdots, 2n,$$

and $\varphi_N(t)$ is therefore a solution of our reduced moment problem (3.10) satisfying the condition (3.11). We pass to the proof of the necessity of the condition of Theorem 3.3. Assume that the reduced moment problem (3.10) admits a solution $\varphi(t)$ satisfying (3.11). We wish to prove that under this condition the quadratic form (3.13) is positive. We first prove that it is non-negative. Introduce a positive number $T$ and set

$$\mu_\nu^{(T)} = \int_{-T}^{T} t^\nu \varphi(t)\,dt, \qquad \nu = 0, 1, 2, \cdots.$$

Let $\{\mu_\nu^{(T)}(L)\}$, $\nu = 0, 1, \cdots, 2n$, be the quantities determined by $\{\mu_\nu^{(T)}\}$ in the same way as $\{\mu_\nu(L)\}$ are determined by $\{\mu_\nu\}$. We obviously have

$$\mu_\nu^{(T)} \to \mu_\nu, \qquad \mu_\nu^{(T)}(L) \to \mu_\nu(L), \qquad \nu = 0, 1, \cdots, 2n, \quad \text{as } T \to \infty.$$

It is clear that for $|z| > T$

$$\frac{1}{L}\int_{-T}^{T} \frac{\varphi(t)\,dt}{z-t} = 1 + \frac{\mu_0^{(T)}}{z} + \frac{\mu_1^{(T)}}{z^2} + \cdots + \frac{\mu_{2n}^{(T)}}{z^{2n+1}} + \cdots,$$

so that

$$f(z) = \exp\left\{\frac{1}{L}\int_{-T}^{T} \frac{\varphi(t)\,dt}{z-t}\right\}$$

is analytic when $z$ is not in the interval $[-T, T]$ and can be expanded in a power series

$$f(z) = 1 + \frac{\mu_0^{(T)}(L)}{z} + \frac{\mu_1^{(T)}(L)}{z^2} + \cdots + \frac{\mu_{2n}^{(T)}(L)}{z^{2n+1}} + \cdots$$

convergent for $|z| > T$. It is easy to prove that

(3.14) $$\Im f(z) \leqq 0, \quad \text{when } \Im z > 0.$$

Indeed, if we write $z = x + iy$, $y > 0$, we have

$$\Im f(z) = |f(z)| \sin\left\{-\frac{1}{L}\int_{-T}^{T} \frac{y\varphi(t)\,dt}{(x-t)^2 + y^2}\right\}.$$

Here, in view of (3.11),

$$0 \leqq \frac{1}{L}\int_{-T}^{T} \frac{y\varphi(t)\,dt}{(x-t)^2+y^2} < \int_{-\infty}^{\infty} \frac{y\,dt}{(x+t)^2+y^2} = \pi,$$

which implies (3.14). An easy application of Lemma 2.2 shows now that there exists an increasing function $\psi(t, T)$ which is constant outside of $[-T, T]$ and which yields the representation

$$f(z) = 1 + \int_{-T}^{T} \frac{d\psi(t, T)}{z - t},$$

whence

$$\mu_\nu^{(T)}(L) = \int_{-T}^{T} t^\nu \, d\psi(t, T), \qquad \nu = 0, 1, 2, \cdots, 2n.$$

This clearly implies that the quadratic form

$$\sum_{i,j=0}^{n} \mu_{i+j}^{(T)}(L) x_i x_j$$

is non-negative, and, on allowing $T \to \infty$, that the quadratic form (3.13) is also non-negative.

To prove that the form (3.13) is positive we use a Lemma due to E. Fischer [1], for the proof of which we refer to Achyeser and Krein [6], also to Fischer, [1].

LEMMA 3.3. *If the quadratic form*

$$\sum_{i,j=0}^{n} \mu_{i+j} x_i x_j$$

*is non-negative and if*

$$\Delta_0 > 0, \cdots, \Delta_{k-1} > 0, \qquad \Delta_k = \cdots = \Delta_n = 0, \qquad k \leqq n,$$

*then there exists a representation (uniquely determined)*

$$\mu_\nu = \sum_{i=1}^{k} \rho_i \xi_i^\nu, \qquad \nu = 0, 1, \cdots, 2n - 1,$$

(3.15)

$$\mu_{2n} = \sum_{i=1}^{k} \rho_i \xi_i^{2n} + M, \qquad M \geqq 0,$$

*where $\xi_1, \cdots, \xi_k$ are real numbers, $\rho_1, \cdots, \rho_k$ are positive and $M = 0$ in case $k = n$.*

Assume now that the form (3.13) is only non-negative. Then, using the representation (3.15), with $\mu_\nu(L)$ instead of $\mu_\nu$, and introducing the polynomials

$$q_k(z) = (z - \xi_1) \cdots (z - \xi_k),$$

$$s_k(z) = q_k(z) + p_k(z) = (z - \eta_1) \cdots (z - \eta_k),$$

analogous to those used in the proof of sufficiency, we construct the function

$$\varphi_{L,k}(t) \equiv \frac{L}{2}\left\{1 - \operatorname{sgn} \frac{s_k(t)}{q_k(t)}\right\},$$

and prove, as before, that

$$\mu_\nu(L) = \int_{-\infty}^{\infty} t^\nu \varphi_{L,k}(t) \, dt, \qquad \nu = 0, 1, \cdots, 2n - 1,$$

with the added relation

$$\mu_{2n}(L) = \int_{-\infty}^{\infty} t^{2n} \varphi_{L,k}(t)\, dt, \quad \text{if} \quad k = n.$$

We then have, whether $k < n$ or $k = n$,

$$\int_{-\infty}^{\infty} [\varphi_{L,k}(t) - \varphi(t)] s_k(t) q_k(t)\, dt = 0.$$

On the other hand, in view of (3.18),

$$\operatorname{sgn}\, [\varphi_{L,k}(t) - \varphi(t)] \operatorname{sgn}\, [s_k(t) q_k(t)] \leqq 0.$$

Therefore we must have almost everywhere $\varphi(t) = \varphi_{L,k}(t)$, which is certainly not possible since $\varphi(t) < L$ while $\varphi_{L,k}(t) = L$ on intervals.

Another theorem of the same kind as Theorem 3.3 can be proved by using practically the same methods as above, with the additional use of Helly's theorem.

THEOREM 3.4. *A necessary and sufficient condition that the moment problem*

$$\mu_\nu = \int_{-\infty}^{\infty} t^\nu \varphi(t)\, dt, \qquad \nu = 0, 1, 2, \cdots,$$

*have a solution $\varphi(t)$ which satisfies the condition*

$$0 \leqq \varphi(t) \leqq L$$

*almost everywhere, is that the sequence $\{\mu_\nu(L)\}$, $\nu = 0, 1, 2, \cdots$, determined by*

$$\exp\left[\frac{1}{L}\left(\frac{\mu_0}{z} + \frac{\mu_1}{z^2} + \cdots\right)\right] \equiv 1 + \frac{\mu_0(L)}{z} + \frac{\mu_1(L)}{z^2} + \cdots$$

*be non-negative in $(-\infty, \infty)$.*

For a proof we refer again to Achyeser and Krein [6].

**6.** We now return to the one-dimensional Hausdorff moment problem and proceed to discuss the remarkable relationship which Hausdorff's solution establishes between the theory of moment sequences and moment functions on one hand and the theory of completely monotonic functions on the other. We shall operate with the following fundamental notions.

A sequence of constants $\{\mu_n\}$, $n = 0, 1, 2, \cdots$, is called *completely monotonic* if all differences

$$\Delta^k \mu_n \equiv \mu_n - \binom{k}{1} \mu_{n+1} + \binom{k}{2} \mu_{n+2} + \cdots + (-1)^k \mu_{n+k} \geqq 0,$$

where $k$, $n$ are any non-negative integers, and $\Delta^0 \mu_n \equiv \mu_n$.

Similarly, a function $f(t)$ is said to be completely monotonic in the interval $(c, \infty)$ if it is defined there and if all differences

$$\Delta_h^k f(t) \equiv f(t) - \binom{k}{1} f(t + h) + \binom{k}{2} f(t + 2h) + \cdots + (-1)^k f(t + kh) \geqq 0,$$

for all non-negative integers $k$ and all $t > c$ and $h > 0$, where

$$\Delta_h^0 f(t) \equiv f(t).$$

The function $f(t)$ is said to be completely monotonic in the interval $[c, \infty)$ if it is completely monotonic in $(c, \infty)$ and if, in addition, $f(c)$ is defined and $f(t)$ is continuous (on the right) at $t = c$, that is, if

$$f(c) = f(c + 0)^*.$$

A sequence $\{\mu_n\}$, $n = 0, 1, 2, \cdots$, or a function $\mu(x)$, $x \geqq 0$, is called respectively a moment sequence or a moment function if there exists a distribution function $\psi(t)$ in $[0, 1]$ such that respectively

$$\mu_n = \int_0^1 t^n \, d\psi(t), \qquad n = 0, 1, 2, \cdots,$$

$$\mu(x) = \int_0^1 t^x \, d\psi(t), \qquad x \geqq 0.$$

A fundamental result of Hausdorff [1] is expressed by

THEOREM 3.4. *The class of functions completely monotonic in* $[0, \infty)$ *is identical with the class of moment functions whose distribution functions are continuous at* $t = 0$.

The proof of this theorem is based on the following facts.

(i) A necessary and sufficient condition that a sequence $\{\mu_n\}$, $n = 0, 1, 2, \cdots$, be a moment sequence is that it be completely monotonic.

This statement is but a rephrasing of the necessary and sufficient condition for the existence of a solution of the moment problem

$$\mu_n = \int_0^1 t^n \, d\psi(t), \qquad n = 0, 1, 2, \cdots,$$

which was proved in Chapter I.

(ii) If $\{\mu_n\}$, $n = 0, 1, 2, \cdots$, is a completely monotonic sequence and $p$ is any positive integer, it is always possible to construct a new sequence $\{\mu_n^{(p)}\}$, $n = 0, 1, 2, \cdots$, by interpolating $(p - 1)$ terms between any two consecutive terms of $\{\mu_n\}$ in such a way that the sequence

$$\mu_n^{(p)}, \qquad \mu_{np}^{(p)} = \mu_n, \qquad n = 0, 1, 2, \cdots,$$

be also completely monotonic. The sequence $\{\mu_n^{(p)}\}$ is uniquely determined.

Let $\{\mu_n\}$, $n = 0, 1, 2, \cdots$, be completely monotonic. Then, by (i), there exists an increasing distribution function such that

$$\mu_n = \int_0^1 t^n \, d\psi, \qquad n = 0, 1, 2, \cdots.$$

* The sequences and functions defined above are completely monotonic; the class of absolutely monotonic sequences and functions is obtained by reversing the order of terms in the differences. In the following we deal only with completely monotonic sequences and functions.

Consider the moment function

$$\mu(x) = \int_0^1 t^x \, d\psi.$$

We obviously have $\mu(n) = \mu_n$. For any $x_0 \geqq 0$ and $h > 0$, the sequence $\{\mu(x_0 + nh)\}$, $n = 0, 1, 2, \cdots$, is also completely monotonic. This follows immediately from the formula

$$\Delta_h^k \mu(x_0 + nh) = \int_0^1 t^{x_0+nh}(1 - t^h)^k \, d\psi \geqq 0.$$

For $x_0 = 0$, $h = 1/p$, we obtain the sequence $\left\{\mu\left(\frac{n}{p}\right)\right\}$ which obviously possesses the properties required in (ii). It remains to prove that the sequence $\{\mu_n^{(p)}\}$ in question is uniquely determined. Assume that $\mu_n^{*(p)}$ is another completely monotonic sequence such that $\mu_{np}^{*(p)} = \mu_n$. By (i), there exists a distribution function $\psi_p(t)$ such that

$$\mu_n^{*(p)} = \int_0^1 u^n \, d\psi_p(u), \qquad n = 0, 1, 2, \cdots,$$

or, by setting $u = t^{1/p}$,

$$\mu_n^{*(p)} = \int_0^1 t^{n/p} \, d\psi_p(t^{1/p}).$$

On substituting here $np$ instead of $n$ and observing that $\mu_{np}^{*(p)} = \mu_n$, we get

$$\int_0^1 t^n \, d\psi_p(t^{1/p}) = \int_0^1 t^n \, d\psi(t), \qquad n = 0, 1, 2, \cdots.$$

Since the moment problem corresponding to the sequence $\{\mu_n\}$ is determined, we conclude that $\psi_p(t^{1/p})$ is substantially equal to $\psi(t)$, so that

$$\mu_n^{*(p)} = \int_0^1 t^{n/p} \, d\psi(t) = \mu\left(\frac{n}{p}\right),$$

which establishes the uniqueness of the interpolated sequence $\{\mu_n^{(p)}\}$.

(iii) If $f(t)$ is completely monotonic in $(c, \infty)$, it is also continuous there*.

This is a well known property which $f(t)$ shares with all monotonic and convex functions.

Finally, we leave to the reader the proof of

(iv) In order that the moment function

$$\mu(x) = \int_0^1 t^x \, d\psi(t)$$

be continuous at $x = 0$ it is necessary and sufficient that $\psi(t)$ be continuous at $t = 0$.

* In fact, $f(t)$ is analytic in $(c, \infty)$, but here we do not need this stronger statement.

We are now ready for the proof of Theorem 3.4. If $\mu(x)$ is any moment function then we know that all sequences $\{\mu(x_0 + hn)\}$, $n = 0, 1, 2, \cdots$, $x_0 \geqq 0$, $h > 0$, are completely monotonic; moreover, if the distribution function $\psi(t)$ of $\mu(x)$ is continuous at $t = 0$ then, by (iv) above, $\mu(x)$ is continuous at $x = 0$; hence $\mu(x)$ is completely monotonic in $[0, \infty)$. It remains to prove the converse statement. Let $\mu(x)$ be a function completely monotonic in $[0, \infty)$. This implies that all sequences $\{\mu(x_0 + nh)\}$, $n = 0, 1, 2, \cdots$, $x_0 \geqq 0$, $h > 0$, are completely monotonic, and, in particular, so is the sequence $\{\mu(n)\}$. Thus, there exists a distribution function $\psi(t)$ such that

$$\mu(n) = \int_0^1 t^n \, d\psi, \qquad n = 0, 1, 2, \cdots . \tag{3.16}$$

The sequences $\left\{\mu\left(\frac{n}{p}\right)\right\}$, $\left\{\int_0^1 t^{n/p} \, d\psi\right\}$, $n = 0, 1, 2, \cdots$, are both interpolating sequences considered in (ii). Therefore they must coincide, so that

$$\mu\left(\frac{n}{p}\right) = \int_0^1 t^{n/p} \, d\psi, \qquad n = 0, 1, 2, \cdots .$$

Thus, the functions $\mu(x)$ and $\int_0^1 t^x \, d\psi(t)$ which are both continuous for $x > 0$ [for $\mu(x)$ this follows from (iii)], coincide for all rational values of $x > 0$; therefore they coincide for all $x > 0$, and

$$\mu(x) = \int_0^1 t^x \, d\psi(t), \qquad x > 0.$$

By (3.16), this equality holds also for $x = 0$, and since, by assumption, $\mu(x)$ is continuous at $x = 0$, $\psi(t)$ must be continuous at $t = 0$, by (iv).

**7.** Theorem 3.4. can be stated in a different form.

THEOREM 3.5. *The class of functions completely monotonic in* $[0, \infty)$ *is identical with the class of functions which are represented by Laplace-Stieltjes integrals*

$$\int_0^\infty e^{-xu} \, d\alpha(u), \tag{3.17}$$

*where* $\alpha(u)$ *is any distribution function in* $[0, \infty)$. *The integral* (3.17) *converges absolutely for all* $x$ *with* $\Re x \geqq 0$.

To each function $\psi(t)$, bounded and increasing in $[0, 1]$ and continuous at $t = 0$, there corresponds a function $\alpha(u)$ bounded and increasing in $[0, \infty)$, by the formula

$$\alpha(u) = \psi(1) - \psi(e^{-u}).$$

Using this fact, it is seen immediately that for all $x \geqq 0$

$$\int_0^1 t^x \, d\psi(t) = \lim_{\delta \to \infty} \int_\delta^1 t^x \, d\psi(t) = \lim_{T \to \infty} \int_0^T e^{-xu} \, d\alpha(u) = \int_0^\infty e^{-xu} \, d\alpha(u).$$

The first proof of Theorems 3.4 and 3.5 is due essentially to Hausdorff [1]. Other proofs were given subsequently by various writers [S. Bernstein, 2; I. J. Schoenberg, 2; J. D. Tamarkin, 1; D. V. Widder, 2].

THEOREM 3.6. *The class of functions completely monotonic in the open interval* $(0, \infty)$ *is identical with the class of functions represented by integrals* (3.17), *where* $\alpha(u)$ *is increasing but not necessarily bounded and such that* (3.17) *converges absolutely for* $\Re x > 0$.

That a function represented by (3.17) is completely monotonic in $(0, \infty)$ is immediately clear from the formula

$$\Delta_h^k f(x) = \int_0^\infty \Delta_h^k(e^{-ux})\, d\alpha(u) = \int_0^\infty e^{-xu}(1 - e^{-hu})^k\, d\alpha(u) \geqq 0.$$

To prove the converse, let $f(x)$ be completely monotonic in $(0, \infty)$. Then $f(x)$ is completely monotonic in $[c, 0)$ where $c > 0$ is arbitrary, and, by Theorem 3.7,

$$f(x) = \int_0^\infty e^{-(x-c)u}\, d\alpha_c(u) = \int_0^\infty e^{-xu} e^{cu}\, d\alpha_c(u), \qquad x \geqq c, \tag{3.18}$$

where $\alpha_c(u)$ is bounded and increasing in $[0, \infty)$, and the integral converges absolutely for $x \geqq c$. We may assume $\alpha_c(0) = 0$. On introducing the function

$$\beta_c(t) = \int_0^t e^{cu}\, d\alpha_c(u)$$

which is increasing but not necessarily bounded in $(0, \infty)$, relation (3.18) can be written in the form

$$f(x) = \int_0^\infty e^{-xu}\, d\beta_c(u), \qquad x \geqq c.$$

For any value of $c_1$, $0 < c_1 < c$, we also have

$$f(x) = \int_0^\infty e^{-xu}\, d\beta_{c_1}(u), \qquad x \geqq c_1,$$

so that, by the uniqueness theorem of the Laplace integral representation, we must have $\beta_c(t) = \beta_{c_1}(t)$ substantially. Thus, there exists an increasing function $\alpha(t)$ to which all $\beta_c(t)$ are substantially equal and which gives the representation (3.17) with all desired properties.

**8.** In the case of a finite interval there exist several methods for obtaining an explicit representation of the solution of the moment problem in terms of the moments. The problem appears to be considerably more difficult in the case of an infinite interval.

For a finite interval, reduced to $(0, 1)$, we start with an elegant solution due to Hausdorff [2]. In dealing with the solution of the moment problem

$$\mu_n = \int_0^1 t^n\, d\psi(t), \qquad n = 0, 1, 2, \cdots, \tag{3.19}$$

there is no loss of generality if $\psi(t)$ is assumed to be normalized so as to satisfy the conditions

$$\text{(3.20)} \qquad \psi(0) = 0, \qquad \psi(t) = \tfrac{1}{2}[\psi(t+0) + \psi(t-0)], \qquad 0 < t < 1.$$

Now consider the sequence of Legendre polynomials

$$L_n(t) = \frac{1}{n!}\frac{d^n}{dt^n}[t^n(1-t)^n], \qquad n = 0, 1, 2, \cdots .$$

These polynomials satisfy the relations

$$\int_0^1 L_i(t)L_j(t)\,dt = \begin{cases} 0, & i \neq j, \\ \dfrac{1}{2j+1}, & i = j, \end{cases}$$

and are related to the classical Legendre polynomials $X_n(t)$ for the interval $(-1, 1)$ by $L_n(t) \equiv X_n(2t-1)$. Put

$$\lambda_n = \mu(L_n) = \int_0^1 L_n(t)\,d\psi, \qquad n = 0, 1, 2, \cdots .$$

It is clear that $\lambda_n$ is a linear combination of the moments $\mu_0, \cdots, \mu_n$, and that, conversely, the moment $\mu_n$ is a linear combination of $\lambda_0, \cdots, \lambda_n$. With this notation we have

THEOREM 3.7. *If the moment problem* (3.19) *has a normalized solution* $\psi(t)$, *then it is represented everywhere in* [0, 1] *by the series*

$$\text{(3.21)} \qquad \psi(t) = \sum_{\nu=0}^{\infty} (2\nu+1)\lambda_\nu \int_0^t L_\nu(u)\,du.$$

It is clear that formula (3.21) holds for $t = 0$ and $t = 1$. We may therefore assume that $t$ is a fixed inner point of the interval (0, 1). From the theory of Legendre series it is known that $\psi(t)$, which is of bounded variation and normalized, can be expanded in the everywhere convergent series

$$\psi(t) = \sum_{\nu=0}^{\infty} (2\nu+1)\lambda_\nu^* L_\nu(t),$$

where

$$\lambda_\nu^* = \int_0^1 L_\nu(u)\psi(u)\,du.$$

Using the relations

$$(2\nu+1)L_\nu(t) = \tfrac{1}{2}(L'_{\nu+1}(t) - L'_{\nu-1}(t)), \qquad \nu \geqq 1,$$

$$\mu(P) = \int_0^1 P(t)\,d\psi(t) = \mu_0 P(1) - \int_0^1 P'(t)\psi(t)\,dt,$$

where $P(t)$ is an arbitrary polynomial, it is readily found that

$$\sum_{\nu=0}^{n} (2\nu+1)\lambda_\nu \int_0^t L_\nu(u)\,du = \sum_{\nu=0}^{n} (2\nu+1)\lambda_\nu^* L_\nu(t) + \tfrac{1}{2}\lambda_{n+1}L_n(t) + \tfrac{1}{2}\lambda_n L_{n+1}(t).$$

On the other hand, since $| L_n(t) | \leqq 1$ in $[0, 1]$,

$$| \lambda_n | = \left| \int_0^1 L_n(t)\, d\psi(t) \right| \leqq \mu_0 ,$$

while $L_n(t) \to 0$, as $n \to \infty$, for every fixed $t$ in $(0, 1)$. Hence

$$\sum_{\nu=0}^{\infty} (2\nu + 1)\lambda_\nu \int_0^t L_\nu(u)\, du = \sum_{\nu=0}^{\infty} (2\nu + 1)\lambda_\nu^* L_\nu(t) = \psi(t),$$

which completes the proof of Theorem 3.7.

It is evident that no changes would be necessary in the above argument if it were assumed only that $\psi(t)$ is of bounded variation, and that an analogous method could be applied in the case where the polynomials $L_n(t)$ are replaced by more general orthogonal polynomials, such as Jacobi polynomials.

An analogous method was used by Widder [4] to obtain an inversion formula for a moment function $\mu(x)$ instead of a moment sequence $\{\mu_n\}$ as discussed above.

THEOREM 3.8. *If a moment function* $\mu(x)$ *admits of the representation*

$$\mu(x) = \int_0^\infty e^{-xt}\, d\alpha(t), \tag{3.22}$$

*where* $\alpha(t)$ *is bounded, increasing and normalized, and* $\alpha(0) = 0$, *then, everywhere in* $[0, \infty)$,

$$\alpha(t) = \sum_{\nu=0}^{\infty} \lambda_\nu \int_0^t L_\nu(u)\, du, \tag{3.23}$$

*where* $L_n(u)$, $n = 0, 1, \cdots$, *is the sequence of Laguerre polynomials defined by*

$$L_n(x) = \sum_{k=0}^{n} \binom{n}{k} \frac{(-x)^n}{k!}, \qquad n = 0, 1, 2, \cdots,$$

*and*

$$\lambda_n \equiv \sum_{k=0}^{n} \binom{n}{k} \frac{\mu^{(k)}(1)}{k!}.$$

Since (3.23) holds for $t = 0$, we may assume $t > 0$. From (3.22) it readily follows that

$$\lambda_n = \int_0^\infty e^{-t} L_n(t)\, d\alpha(t), \qquad n = 0, 1, 2, \cdots.$$

On the other hand, from the theory of Laguerre series it is known that

$$\alpha(t) = \sum_{\nu=0}^{\infty} \lambda_\nu^* L_\nu(t),$$

where

$$\lambda_\nu^* = \int_0^\infty e^{-u} L_\nu(u)\alpha(u)\, du.$$

Now, using the formula

$$\int_0^t L_\nu(u)\,du = L_\nu(t) - L_{\nu+1}(t), \qquad \nu = 0, 1, 2, \cdots,$$

we have

$$\lambda_0 = \int_0^\infty e^{-t} L_0(t)\alpha(t)\,dt = \lambda_0^*,$$

while, for $\nu \geqq 1$,

$$\lambda_\nu - \lambda_{\nu-1} = \int_0^\infty e^{-t} L_\nu(t)\alpha(t)\,dt = \lambda_\nu^*.$$

Hence

$$\sum_{\nu=0}^n \lambda_\nu \int_0^t L_\nu(u)\,du = \sum_{\nu=0}^n \lambda_\nu^* L_\nu(t) - \lambda_n L_{n+1}(t).$$

The conclusion of Theorem 3.8 now follows if we observe that

$$\lambda_n = O(1), \qquad L_{n+1}(t) = O(n^{-1/4}), \qquad \text{as } n \to \infty.$$

The above proof may be applied if $\alpha(t)$ is to be of bounded variation over $[0, \infty)$ and the polynomials $L_n(t)$ are replaced by generalized Laguerre polynomials.

As observed by Hille [1], Theorems 3.7 and 3.8 may be considered as special cases of a more general inversion formula for the Laplace integral

$$f(z) = \int_0^\infty e^{-zu} F(u)\,du,$$

which is obtained whenever we have a generalized expansion

$$e^{-zu} = \sum_{n=1}^\infty \Phi_n(u)\Psi_n(z).$$

**9.** There are several methods for obtaining the solution of the moment problem (3.19) as a limit of a convergent sequence of step-functions having a finite or an infinite number of steps. Such is, in the first place, the method used in Chapter II, where the solution $\psi(t)$ was obtained as the limit of a sequence of distribution functions $\psi_n(t)$ of finite orders. These distribution functions were characterized by the condition that they have their first $(2n + 1)$ moments the same as $\psi(t)$ itself. Here we mention different methods due to Hausdorff [2] and to Widder [3], where the approximating function is not required to have the same initial moments as the solution $\psi(t)$.

In the theory of Hausdorff the quantities

$$\lambda_{n\nu} \equiv \frac{n!}{\nu!(n-\nu)!}\Delta^{n-\nu}\mu_\nu = \binom{n}{\nu}\Delta^{n-\nu}\mu_\nu$$

play an important role. We observe that, if $\psi(t)$ is a solution of the moment problem (3.19), then

$$(n+1)\lambda_{n\nu} = \frac{(n+1)!}{\nu!(n-\nu)!}\int_0^1 t^\nu(1-t)^{n-\nu}\,d\psi$$

appears as a certain weighted average of the distribution characterized by $\psi(t)$.

Hausdorff introduces the normalized step function $\chi_n(t)$ defined as follows:

$$\chi_n(0) = 0, \qquad \chi_n(1) = \sum_{\nu=0}^{n} \lambda_{n\nu} = \mu_0,$$

and, for $0 < t < 1$,

$$\chi_n(t-0) = \sum_{\nu<nt} \lambda_{n\nu}, \qquad \chi_n(t+0) = \sum_{\nu\leq nt} \lambda_{n\nu},$$

$$\chi_n(t) = \tfrac{1}{2}[\chi_n(t+0) + \chi_n(t-0)].$$

He proves that $\chi_n(t)$ converges to the solution $\psi(t)$ of the moment problem (3.19) for all $t$ in $[0, 1]$.

To explain the method used by Widder, assume first that the moment problem (3.19) has an absolutely continuous solution so that $d\psi(t) = \varphi(t)dt$, and hence

$$(n+1)\lambda_{n\nu} = \frac{(n+1)!}{\nu!(n-\nu)!}\int_0^1 t^\nu(1-t)^{n-\nu}\varphi(t)\,dt.$$

The maximum of the factor $t^\nu(1-t)^{n-\nu}$ is attained at $t = \frac{\nu}{n}$. Thus, if we fix the value $t_0$ in $(0, 1)$ and select $\nu$ as function of $n$ and $t_0$ so that $\frac{\nu}{n} = t_0$ approximately, we may expect that $(n+1)\,\lambda_{n\nu} \to \varphi(t_0)$, as $n \to \infty$. Changing notation, replace $n - \nu$ by $k$. We then get the expression

$$\frac{(\nu+k+1)!}{\nu!k!}\Delta^k\mu_\nu,$$

where $\frac{\nu}{k+\nu}$ must approximately equal $t_0$. This requirement is satisfied if we take $\nu = \left[\frac{kt}{1-t}\right]$. We thus naturally obtain Widder's operator*

$$L_{k,t}(\mu) \equiv \frac{(\nu+k+1)!}{\nu!k!}\Delta^k\mu_\nu, \qquad \nu = \left[\frac{kt}{1-t}\right].$$

* Widder introduced this operator by analogy with the operator

$$L_{kt}[f] \equiv \frac{(-1)^k}{k!} f^{(k)}\left(\frac{k}{t}\right)\left(\frac{k}{t}\right)^{k+1}$$

which he used in the inversion theory of the Laplace-Stieltjes integral $f(x) = \int_0^\infty e^{-xt}\,d\alpha(t)$. The first idea of using such an operator for this purpose is apparently due to Stieltjes [6, 2 (383)].

We now state some of the results of Widder, referring for the proofs to Widder [3].

Let

$$L_{k,t}(\mu) = \frac{(\nu + k + 1)!}{\nu! k!} \Delta^k \mu_\nu, \qquad \nu = \left[\frac{kt}{1-t}\right],$$

$$S_{k,t}(\mu) = -\mu_\infty - \sum_{i=\nu+1}^{\infty} \frac{(i+k)!}{i! k!} \Delta^{k+1} \mu_i .$$

Then:

(i) if

$$\mu_n = \int_0^1 t^n \, d\psi, \qquad n = 0, 1, 2, \cdots,$$

where $\psi(t)$ is of bounded variation in [0, 1] and is normalized by the conditions

$$\psi(1) = 0, \qquad \psi(t) = \tfrac{1}{2}[\psi(t+0) + \psi(t-0)], \qquad 0 < t < 1,$$

we have

$$\lim_{k \to \infty} S_{k,t}(\mu) = \psi(t), \qquad 0 < t < 1;$$

(ii) if

$$\mu_n = \int_0^1 t^n \varphi(t) \, dt, \qquad n = 0, 1, 2, \cdots,$$

where $\varphi(t)$ is integrable in (0, 1), we have

$$\lim_{k \to \infty} L_{k,t}(\mu) = \varphi(t)$$

almost everywhere in (0, 1).

On introducing the operator

$$L^*_{k,t}(\mu) \equiv \frac{\Gamma(\omega + k + 2)}{\Gamma(\omega + 1)\Gamma(k+1)} \Delta^k \mu_\omega, \qquad \omega = \frac{kt}{1-t},$$

where

$$\Delta^k \mu_\omega \equiv \int_0^1 u^\omega (1-u)^k \varphi(u) \, du,$$

Widder proves further that if

$$\mu_n = \int_0^1 u^n \varphi(u) \, du, \qquad n = 0, 1, 2, \cdots,$$

where $\varphi(u)$ is analytic in the circle $| u - \frac{1}{2} | < \frac{1}{2}$, then for any $t$ in this circle

$$\lim_{k \to \infty} L^*_{k,t}(\mu) = \varphi(t).$$

We also mention various results of Widder concerning the number of changes of sign in the sequences $\{\Delta^k \mu_n\}$, $n = 0, 1, 2, \cdots$, in their relation to the number of changes of trend of $\psi(t)$.

**10.** In the case of the Stieltjes moment problem a formal solution was given by LeRoy [2]. It was put on a rigorous basis by Hardy [$1_b$]. (For a slightly different treatment see Titchmarsh, [1]).

Suppose that the moment problem

$$\mu_n = \int_0^\infty t^n \varphi(t)\, dt, \qquad n = 0, 1, 2, \cdots$$

has a solution $\varphi(t)$ such that

$$\int_0^\infty |\varphi(t)|\, e^{k\sqrt{t}}\, dt < \infty \tag{3.24}$$

for a certain $k > 0$ (in the case where $\varphi(t) \geqq 0$ we know, by Corollary 1.1, that the problem is determined). It is readily seen that the function

$$g(z) = \int_0^\infty \varphi(t) J_0\{2\sqrt{tz}\}\, dt \tag{3.25}$$

is analytic for $z > 0$ and is represented by the power series expansion

$$g(z) = \sum_{n=0}^\infty \frac{(-1)^n \mu_n}{(n!)^2} z^n,$$

for $|z|$ sufficiently small. This follows from (3.24) and from the expansion for the Bessel function

$$J_0(2\sqrt{z}) = \sum_{n=0}^\infty \frac{(-1)^n z^n}{(n!)^2}.$$

Thus, to find $\varphi(t)$, it remains to solve the integral equation (3.25). It turns out that almost everywhere

$$\varphi(t) = \int_0^\infty J_0\{2\sqrt{ty}\} g(y)\, dy,$$

where the integral on the right in general does not converge, but must be evaluated by some of the known summability methods, as, for instance,

$$\lim_{\delta \to 0} \int_0^\infty e^{-\delta y} J_0\{2\sqrt{ty}\} g(y)\, dy,$$

or

$$\lim_{Y \to \infty} \frac{1}{Y} \int_0^Y (Y - y) J_0\{2\sqrt{ty}\} g(y)\, dy.$$

Another method proposed by Hardy [$1_a$] is to construct the series

$$h(z) = \sum_{n=0}^{\infty} \frac{(-1)^n \mu_n}{(2n)!} z^n = \int_0^{\infty} \varphi(t) \cos \{\sqrt{zt}\}\, dt,$$

so that

$$H(u) \equiv \int_0^{\infty} e^{-z} h\left(\frac{z^2}{u}\right) dz = \int_0^{\infty} \varphi(t)\, dt \int_0^{\infty} e^{-z} \cos\left\{z\left(\frac{t}{u}\right)^{1/2}\right\} dz$$

$$= \int_0^{\infty} \frac{u\varphi(t)\, dt}{t+u},$$

and to find $\varphi(t)$ by applying the inversion formula, [Introduction, **5**] to the equation

$$\frac{H(u)}{u} = \int_0^{\infty} \frac{\varphi(t)\, dt}{t+u}.$$

In case $\varphi(t)$ satisfies the condition

$$\int_0^{\infty} |\varphi(t)|\, e^{kt}\, dt < \infty, \qquad k > 0,$$

instead of (3.24), the problem of finding $\varphi(t)$ reduces to that of inverting a Laplace integral, for we may put

$$g(z) = \sum_{n=0}^{\infty} \frac{(-1)^n \mu_n}{n!} z^n = \int_0^{\infty} \varphi(t) e^{-zt}\, dt.$$

(See also Titchmarsh [1]).

**11.** In the present section we discuss various moment-problems which differ from those which we have discussed on preceding pages by the removal of the requirement that the solution shall be an increasing function. (We have met such modified problems in some isolated instances before). Thus we shall deal with moment problems of the type

$$\mu_n = \int_0^1 t^n\, d\psi, \qquad n = 0, 1, 2, \cdots, \tag{3.26}$$

where $\psi(t)$, instead of being assumed increasing, is now assumed to belong to more or less restricted sub-classes of the general class of functions of bounded variation. An important fact is revealed by

THEOREM 3.9. *A solution of bounded variation of the moment problem* (3.26), *if it exists at all, is substantially unique.*

Indeed, assuming that (3.26) has two solutions $\psi_1(t)$, $\psi_2(t)$, of bounded variation, we may write

$$\psi_1(t) = \psi_1'(t) - \psi_1''(t), \qquad \psi_2(t) = \psi_2'(t) - \psi_2''(t),$$

where $\psi_1'(t)$, $\psi_1''(t)$, $\psi_2'(t)$, $\psi_2''(t)$ are bounded increasing functions. Our assumption yields, for $n = 0, 1, 2, \cdots$,

$$\int_0^1 t^n\, d(\psi_1'(t) + \psi_2''(t)) = \int_0^1 t^n\, d(\psi_2'(t) + \psi_1''(t)).$$

Thus, two bounded increasing functions $\psi_1'(t) + \psi_2''(t)$, $\psi_2'(t) + \psi_1''(t)$ have the same moments of all orders over the finite interval (0, 1) and therefore must be substantially equal, which, in turn, implies the substantial equality of the functions $\psi_1(t)$ and $\psi_2(t)$.

Introduce again the quantities

$$\lambda_{n\nu} \equiv \binom{n}{\nu} \Delta^{n-\nu} \mu_\nu = \binom{n}{\nu} \int_0^1 t^\nu (1 - t)^{n-\nu}\, d\psi = \mu(\lambda_{n\nu}(t)), \tag{3.27}$$

where

$$\lambda_{n\nu}(t) \equiv \binom{n}{\nu} t^\nu (1 - t)^{n-\nu}. \tag{3.28}$$

We start with a lemma [Hausdorff, 2].

LEMMA 3.4. *If $p$ is any number $\geqq 1$, then the expression*

$$(n + 1)^{p-1} \sum_{\nu=0}^{n} | \lambda_{n\nu} |^p$$

*is an increasing function of $n$.*

Since

$$(n + 1)\lambda_{n\nu}(t) = (n + 1 - \nu)\lambda_{n+1,\nu}(t) + (\nu + 1)\lambda_{n+1,\nu+1}(t),$$

we have

$$(n + 1)\lambda_{n\nu} = (n + 1 - \nu)\lambda_{n+1,\nu} + (\nu + 1)\lambda_{n+1,\nu+1},$$

which can be written in the form

$$\frac{n+1}{n+2}\lambda_{n\nu} = \frac{n+1-\nu}{n+2}\lambda_{n+1,\nu} + \frac{\nu+1}{n+2}\lambda_{n+1,\nu+1}$$

and gives, by an easy application of Hölder's inequality,

$$\left(\frac{n+1}{n+2}\right)^p | \lambda_{n\nu} |^p \leqq \frac{n+1-\nu}{n+2} | \lambda_{n+1,\nu} |^p + \frac{\nu+1}{n+2} | \lambda_{n+1,\nu+1} |^p,$$

from which Lemma 3.4 is easily derived if we sum over $\nu$ from 0 to $n + 1$.

It follows that $(n + 1)^{1-(1/p)} \left( \sum_{\nu=0}^{n} | \lambda_{n\nu} |^p \right)^{1/p}$ is increasing when $n$ increases, and this holds true also in the limiting case $p = \infty$ if we replace the expression in question by $(n + 1) \max_\nu | \lambda_{n\nu} |$. Thus,

$$\lim_{n\to\infty} (n + 1)^{1-(1/p)} \left( \sum_{\nu=0}^{n} | \lambda_n |^p \right)^{1/p} \quad \text{or} \quad \lim_{n\to\infty} \{(n + 1) \max_\nu | \lambda_{n\nu} |\}$$

exists (as a finite or infinite number) for $1 \leqq p \leqq \infty$; it will be denoted by $M_p$.

We may now state our main theorem.

THEOREM 3.10. *A necessary and sufficient condition that the moment-problem* (3.26) *have*

(i) *a solution* $\psi(t)$ *of bounded variation, is that* $M_1 < \infty$.

(ii) *a solution* $\psi(t) = \int_0^t \varphi(u)\,du$, *where* $\varphi(t) \in L_p(0, 1)$, $p > 1$, *is that* $M_p < \infty$.

(iii) *a solution* $\psi(t) = \int_0^t \varphi(u)\,du$, *where* $\varphi(t)$ *is bounded, is that* $M_\infty < \infty$.

*Furthermore, if the conditions of cases* (i, ii, iii) *are satisfied, then we have respectively:*

(i) $\int_0^1 | d\psi(t) | = M_1$, *provided* $\psi(t)$ *is normalized,*

(ii) $\left(\int_0^1 | \varphi(t) |^p \,dt\right)^{1/p} = M_p$,

(iii) $\operatorname*{ess\,sup}_{0 \leq t \leq 1} | \varphi(t) | = M_\infty$.

*In all cases the solution is unique.*

We start with the case (i). If problem (3.26) admits of a solution $\psi(t)$ of bounded variation, then

$$\sum_{\nu=0}^{n} | \lambda_{n\nu} | = \sum_{\nu=0}^{n} \binom{n}{\nu} \left| \int_0^1 t^\nu (1 - t)^{n-\nu} \,d\psi(t) \right|$$

$$\leq \sum_{\nu=0}^{n} \binom{n}{\nu} \int_0^1 t^\nu (1 - t)^{n-\nu} \,| d\psi(t) | = \int_0^1 | d\psi(t) |,$$

so that

$$M_1 = \lim_n \sum_{\nu=0}^{n} | \lambda_{n\nu} | \leq \int_0^1 | d\psi(t) | < \infty,$$

which proves the necessity of the condition of the theorem. Assume now that this condition is satisfied. Then, if $P(t)$ is an arbitrary polynomial, of degree $k$, and if $B_n(t, P)$ is the corresponding Bernstein polynomial, of degree $n$, we have [1.2]

$$P(t) = B_n(t; P) + \epsilon_n, \qquad \epsilon_n = \sum_{s=1}^{n} \frac{p_{ks}(t)}{n^s},$$

so that

$$\mu(P) = \mu(B_n) + \mu(\epsilon_n),$$

$$| \mu(P) | \leq \sum_{\nu=0}^{n} \left| P\left(\frac{\nu}{n}\right) \right| \lambda_{n\nu} + | \mu(\epsilon_n) | \leq \| P \|_C M_1 + O(1/n),$$

where

$$\| P \|_C = \max_{0 \leq t \leq 1} | P(t) |.$$

On allowing $n \to \infty$, we get

$$| \mu(P) | \leqq \| P \|_C M_1 ,$$

which shows that $\mu(P)$ is a linear functional defined on the linear subspace $\mathfrak{M}_0$ (the manifold of all polynomials) of the space $C$ of functions continuous in [0, 1] and that the norm of $\mu(P)$ does not exceed $M_1$. By Introduction, **4**, $\mu(P)$ can be extended over the whole space $C$ so as to remain linear and with preservation of the norm. But it is well known [Banach, 1] that the general form of such a functional is

$$\mu(X) = \int_0^1 X(t)\, d\psi(t), \tag{3.29}$$

where $X \equiv X(t)$ is any element of the space $C$ and $\psi(t)$ is a function of bounded variation. Furthermore, it is known that if $\psi(t)$ is normalized by the condition

$$\psi(t) = \tfrac{1}{2}[\psi(t + 0) + \psi(t - 0)], \qquad 0 < t < 1,$$

(which does not change the value of the integral in (3.23)), then the norm of $\mu(X)$ will be precisely $\int_0^1 | d\psi(t) |$. Thus

$$\int_0^1 | d\psi(t) | \leqq M_1 ,$$

and since clearly

$$\int_0^1 t^n\, d\psi(t) = \mu_n , \qquad n = 0, 1, 2, \cdots ,$$

it follows that $\psi(t)$ is the solution of bounded variation of the moment problem (3.26), and that $\int_0^1 | d\psi(t) | = M_1$.

In the case (ii) let $1 < p < \infty$ and let

$$d\psi(t) = \varphi(t)\, dt, \qquad \varphi \in L_p(0, 1),$$

be a solution of the moment problem (3.26). Then, using the notation (3.27), (3.28), we have, on writing $p' = \dfrac{p}{p-1}$,

$$| \lambda_{n\nu} | = \left| \int_0^1 \lambda_{n\nu}(t)\varphi(t)\, dt \right| \leqq \left( \int_0^1 \lambda_{n\nu}(t) | \varphi(t) |^p\, dt \right)^{1/p} \left( \int_0^1 \lambda_{n\nu}(t)\, dt \right)^{1/p'}$$

$$= \left( \int_0^1 \lambda_{n\nu}(t) | \varphi(t) |^p\, dt \right)^{1/p} \left( \frac{1}{n+1} \right)^{1/p'},$$

since

$$\int_0^1 \lambda_{n\nu}(t)\, dt = \binom{n}{\nu} \int_0^1 t^\nu (1 - t)^{n-\nu}\, dt = \binom{n}{\nu} \frac{\Gamma(\nu + 1)\Gamma(n - \nu + 1)}{\Gamma(n + 2)} = \frac{1}{n+1}.$$

Thus, on writing

$$\|\varphi\|_p = \left(\int_0^1 |\varphi(t)|^p\,dt\right)^{1/p},$$

we have

$$(n+1)^{p-1}\sum_{\nu=0}^{n}|\lambda_{n\nu}|^p \leqq \int_0^1 \sum_{\nu=0}^{n}\binom{n}{\nu}t^\nu(1-t)^{n-\nu}|\varphi(t)|^p\,dt = \|\varphi\|_p^p,$$

and finally,

$$(n+1)^{1-1/p}\left(\sum_{\nu=0}^{n}|\lambda_{n\nu}|^p\right)^{1/p} \leqq M_p \leqq \|\varphi\|_p,$$

which shows the necessity of the condition of the theorem in the case (ii).

Assume now that this condition is satisfied. Then, with the notation used in the case (i), we have

$$|\mu(B_n)| = \left|\sum_{\nu=0}^{n}P\left(\frac{\nu}{n}\right)\lambda_{n\nu}\right| \leqq M_p\left\{\frac{1}{n+1}\sum_{\nu=0}^{n}\left|P\left(\frac{\nu}{n}\right)\right|^{p'}\right\}^{1/p'}$$

Since $B_n(t; P) \to P(t)$ uniformly in $[0, 1]$, as $n \to \infty$, and since the expression in brackets in the right-hand member of the preceding inequality tends to $\left\{\int_0^1 |P(t)|^{p'}\,dt\right\}^{1/p'}$, we see that

$$|\mu(P)| \leqq M_p\,\|P\|_{p'}$$

so that $\mu(P)$ appears as a linear functional defined on the linear subspace $\mathfrak{M}_0$ of the space $L_{p'}(0, 1)$. Using the fact that the general linear functional on $L_{p'}(0, 1)$ is given by

$$\int_0^1 X(t)\varphi(t)\,dt, \qquad \varphi \in L_p(0, 1),$$

an argument analogous to that used in the case (i) readily shows that

$$\mu(P) = \int_0^1 P(t)\varphi(t)\,dt,$$

whence

$$\mu_n = \int_0^1 t^n\varphi(t)\,dt, \qquad n = 0, 1, 2, \cdots.$$

The argument to be used in the case (iii), which is the limiting case of (ii), as $p \to \infty$, is quite analogous to that used in the case (ii), and may be left to the reader.

The case (i) of Theorem 3.10 can also be restated in the following form:

*A necessary and sufficient condition that the sequence* $\{\mu_n\}$, $n = 0, 1, 2, \cdots$, *be a sequence of moments* [*over the interval* (0, 1)] *of a function of bounded variation, is that it be a difference of two completely monotonic sequences.*

The argument used in proving Theorem 3.10 in the case (i) can be readily extended to any number of dimensions. Thus we may state the following result (using the notation of Chapter I):

*A necessary and sufficient condition that the moment problem*

$$\mu_{i,j} = \int_0^1 \int_0^1 u^i v^j \, d\Phi$$

*have a solution of bounded variation is the existence of a fixed number M such that*

$$\sum_{i=0}^{n} \sum_{j=0}^{m} \binom{n}{i}\binom{m}{j} \left| \Delta_1^{n-i} \Delta_2^{m-j} \mu_{ij} \right| \leqq M,$$

for all values $\geqq 0$ of the integers $m, n$.

**12.** There are other methods of obtaining criteria for the existence of a solution of the moment problem (3.26). They consist in applying various summability methods to the explicit representation (3.21) of the solution and then proceeding along the lines familiar from the theory of orthogonal expansions [Kaczmarz und Steinhaus, 1]. We refer to these sources and also to [Hausdorff, 1] for proofs of statements which follow.

Let the matrix

$$T = \begin{pmatrix} b_{11} & b_{12} & \cdots & b_{1m_1} \\ b_{21} & b_{22} & \cdots & b_{2m_2} \\ \cdots & \cdots & \cdots & \cdots \\ b_{n1} & b_{n2} & \cdots & b_{nm_n} \\ \cdots & \cdots & \cdots & \cdots \end{pmatrix}$$

correspond to a regular definition of summability, and let

$$\Phi_n(t, u) = \sum_{k=0}^{m_n} b_{nk} \frac{2k+1}{2} L_k(t) L_k(u) = \Phi_n(u, t)$$

be the $n$-th transform of the $n$-th Legendre kernel

$$\sum_{\nu=0}^{n} \frac{2\nu+1}{2} L_\nu(t) L_\nu(u).$$

It is assumed that there exists a fixed constant $A$* such that

$$\int_0^1 |\Phi_n(t, u)| \, du \leqq A, \qquad 0 \leqq t \leqq 1, \qquad n = 0, 1, 2, \cdots.$$

Let

$$\varphi_n(u) = \mu_t\{\Phi_n(t, u)\},$$

* This assumption is automatically satisfied when $\Phi_n(t, k) \geqq 0$, which includes cases treated by Hausdorff, when the transformation $T$ is $(C, 2)$ or a de la Vallée-Poussin transformation.

where $\mu_t$ means that the functional $\mu$ is taken relative to the variable $t$. Then a necessary and sufficient condition for the moment problem (3.26) to have a solution $\psi(t)$ of one of the following types:

(i) bounded and increasing,
(ii) of bounded variation
(iii) of the form $\psi(t) = \int_0^t \varphi(u)\,du,\ \varphi \in L_p(0, 1),\ p > 1,$
(iv) of the same form as in (iii), with $\varphi(u)$ essentially bounded,
(v) of the same form as in (iii), with $\varphi \in L_1(0, 1)$,

is respectively:

(i) $\varphi_n(u) \geqq 0, \qquad 0 \leqq u \leqq 1, \qquad n = 0, 1, 2, \cdots;$
(ii) $\int_0^1 |\varphi_n(u)|\,du \leqq M, \qquad n = 0, 1, 2, \cdots;$
(iii) $\|\varphi_n\|_p \leqq M, \qquad n = 0, 1, 2, \cdots;$
(iv) $\operatorname{ess\,sup}_{0 \leqq u \leqq 1} |\varphi_n(u)| \leqq M, \qquad n = 0, 1, 2, \cdots;$
(v) $\|\varphi_n(u) - \varphi_m(u)\|_1 \to 0, \qquad \text{as} \quad n, m \to \infty.$

Here $M$ denotes a fixed constant.

**13.** The situation in the case of an infinite interval is essentially different from that of a finite interval, as is shown by the following theorem [Boas, 3]

THEOREM 3.11. *Given an arbitrary sequence of real constants,* $\{\mu_n\}$, $n = 0, 1, \cdots$, *there exist infinitely many functions* $\psi(t)$ *of bounded variation over* $(-\infty, \infty)$, *for which*

$$\mu_n = \int_{-\infty}^{\infty} t^n\,d\psi, \qquad n = 0, 1, 2, \cdots.$$

*The same is true for the Stieltjes moment problem.*

Following the idea of Boas, we give a proof for the first (Hamburger) part of the theorem only. The modifications necessary for the second (Stieltjes) part are obvious. Theorem 3.11 will be proved if we can show that an arbitrary given sequence $\{\mu_n\}$, $n = 0, 1, 2, \cdots$, can be represented in infinitely many ways in the form $\mu_n = \mu_n' - \mu_n''$, where the constants $\mu_n'$ and $\mu_n''$ are such that all determinants

$$\Delta_n' = |\mu_{i+j}'|_{i,j=0}^n, \qquad \Delta_n'' = |\mu_{i+j}''|_{i,j=0}^n, \qquad n = 0, 1, 2, \cdots,$$

are positive. This statement is clear for $n = 0$. Assume it proved for $0, 1, 2, \cdots, n-1$, and prove its validity for $n$. Thus assume that the constants $\mu_0', \cdots, \mu_{2n-2}'$ and $\mu_0'', \cdots, \mu_{2n-2}''$ have been selected in such a way that

$$\mu_\nu' - \mu_\nu'' = \mu_\nu, \qquad \nu = 0, 1, \cdots, 2n-2,$$

$$\Delta_\nu' > 0, \qquad \Delta_\nu'' > 0, \qquad \nu = 0, 1, \cdots, n-1.$$

Now select $\mu_{2n-1}'$, $\mu_{2n-1}''$ arbitrarily, subject only to the requirement that $\mu_{2n-1}' - \mu_{2n-1}'' = \mu_{2n-1}$. We have to select $\mu_{2n}'$ and $\mu_{2n}''$ so that

$$\Delta_n' > 0, \qquad \Delta_n'' > 0.$$

This is always possible if we take $\mu'_{2n}$, $\mu''_{2n}$ sufficiently large since they enter in the first degree in the expansions of $\Delta'_n$, $\Delta''_n$, with coefficients $\Delta'_{n-1}$, $\Delta''_{n-1}$, respectively, which, by assumption, are positive. It remains to select $\mu'_{2n}$, $\mu''_{2n}$ in such a way that $\mu'_{2n} - \mu''_{2n} = \mu_{2n}$, which is always possible.

Pólya [3] gave an entirely different proof of Boas' theorem, which enabled him also to show that the solution can be made to be either (i) a step-function with the points of increase at an arbitrarily prescribed sequence of distinct real numbers $\{x_\nu\}$, $\nu = 1, 2, \cdots$, with no finite limit point, or (ii) an entire transcendental function. Pólya also points out that the case (i) is essentially contained in an older result of Borel [1].

**14.** Another modification of the classical moment problem consists in replacing the function $t^n$ by $t^{k_n}$. In the paper [1] Hausdorff extends his methods to this more general situation. Let

$$k_0 = 0 < k_1 < k_2 < \cdots < k_n < \cdots \tag{3.30}$$

be a given sequence of real numbers. Let

$$\Omega_n(t) = (t - k_0)(t - k_1) \cdots (t - k_n), \qquad \Omega_{-1}(t) \equiv 1.$$

Hausdorff considers the moment problem

$$\mu_n = \int_0^1 t^{k_n}\, d\psi(t), \qquad n = 0, 1, 2, \cdots, \tag{3.31}$$

where $\psi(t)$ is a bounded increasing function. He introduces the quantities

$$\lambda_{n\nu} = \frac{\Omega'_n(k_0)}{\Omega'_\nu(k_0)} \sum_{i=\nu}^{n} \frac{\mu_i}{\Omega'_n(k_i)},$$

which reduce to those discussed above in the case $k_0 = 0$, $k_1 = 1, \cdots$, $k_n = n, \cdots$. The sequence $\{\mu_n\}$ is called completely monotonic relative to the sequence $\{k_n\}$ if

$$\lambda_{n\nu} \geqq 0, \qquad n = 0, 1, 2, \cdots; \qquad \nu = 0, 1, \cdots, n.$$

[A sequence completely monotonic relative to the sequence $\{0, 1, 2, \cdots\}$ is what we have previously called simply a completely monotonic sequence]. It is established that the existence of a fixed constant $A$ such that

$$\sum_{\nu=0}^{n} |\lambda_{n\nu}| \leqq A$$

is a necessary and sufficient condition in order that the sequence $\{\mu_n\}$ be the difference of two sequences completely monotonic relative to $\{k_n\}$. If, in addition, it is assumed that

$$\sum_{n=1}^{\infty} 1/k_n = \infty \tag{3.32}$$

then it is proved that a necessary and sufficient condition for the existence of a solution of the problem (3.31) is that the sequence $\{\mu_n\}$ be completely monotonic relative to $\{k_n\}$, in other words, that $\lambda_{n\nu} \geqq 0$. It is also proved that, under the condition (3.32), the problem (3.31) is always determined.

Hausdorff's investigations have been continued by Hallenbach [1] in the case when the series $\sum_{n=1}^{\infty} \frac{1}{k_n} < \infty$. In this case the condition of being completely monotonic is necessary but not sufficient for the existence of a solution of (3.31). There exists, however, another condition which is equivalent to that of being completely monotonic in the case (3.32), and which turns out to be both necessary and sufficient if we assume only (3.30). Consider all "polynomials"

$$P_n(t) = x_0 t^{k_0} + \cdots + x_n t^{k_n}$$

and the functional

$$\mu(P_n) \equiv x_0\mu_0 + \cdots + x_n\mu_n .$$

This functional is said to be positive (non-negative) if the condition $P_n(t) \geqq 0$ in [0, 1] and $P_n(t) \not\equiv 0$ ($P_n(t) \geqq 0$ in [0, 1]) implies $\mu(P_n) > 0$ ($\mu(P_n) \geqq 0$). On introducing this notion, it is proved by a simple modification of methods used in Chapter I, that a necessary and sufficient condition for the existence of a solution of the moment problem (3.31) is that the functional $\mu(P_n)$ be non-negative.

Some of the methods used in Chapter II in treating the question of determinedness can be also extended to the present case. Let, for a fixed $t_0$ in [0, 1],

$$\rho_n(t_0) \equiv \min_{P_n} \mu(P_n),$$

where the minimum is taken over all polynomials $P_n(t) \geqq 0$ in [0, 1] and such that $P_n(t_0) = 1$. It is proved that a necessary and sufficient condition for the modified moment problem (3.31) to be indeterminate is that there exist a subinterval of the interval [0, 1] and a positive number $\delta$ such that, in this subinterval,

$$\rho(t) = \lim_{n\to\infty} \rho_n(t) \geqq \delta > 0.$$

Hallenbach also extends several results of Hausdorff to the case of two dimensions. Hausdorff's problems for one and two dimensions are also treated in Hildebrandt and Schoenberg [1]. There they appear as special cases of a more general theory of certain linear equations.

The case of a modified Stieltjes moment problem

$$\mu_n = \int_0^{\infty} t^{k_n}\, d\psi(t), \qquad n = 0, 1, 2, \cdots ,$$

has been treated recently by Boas [2].

# CHAPTER IV

## APPROXIMATE QUADRATURES.

**1.** This chapter is devoted to a detailed study of the approximate quadrature formula (II,**8**)

$$\int_{-\infty}^{\infty} f(t)\,d\psi \sim \sum_{j=0}^{n} \rho_n(x_j)f(x_j), \tag{4.1}$$

where $\psi(t)$ is any solution of the moment problem (2.1), the abscissas

$$x_j \equiv x_{j,n}, \qquad x_{0n} < x_{1n} < \cdots < x_{nn}$$

are the roots of the quasi-orthogonal polynomial

$$q_{n+1}(z) = \omega_{n+1}(z) + A_n\omega_n(z), \tag{4.2}$$

$A_n$ is constant, and the coefficients

$$\rho_n(x_j) \equiv \rho_j \equiv \rho_{jn}, \qquad j = 0, 1, 2, \cdots, n,$$

are given (cf. II, **8, 9, 10**) by

$$\rho_j = \int_{-\infty}^{\infty} \frac{q_{n+1}(t)\,d\psi}{(t - x_{jn})q'_{n+1}(x_{jn})} = \rho_n(x_{jn}) = \frac{1}{K_n(x_{jn})}. \tag{4.3}$$

The function $f(t)$ is assumed to be finite for all $t$ and integrable with respect to the function $\psi(t)$ in question.

We rewrite (4.1) as

$$\begin{aligned} I_\psi(f) &\equiv \int_{-\infty}^{\infty} f(t)\,d\psi = \sum_{j=0}^{n} \rho_j f(x_j) + R_{n,\psi}(f) \\ &= Q_n(f) + R_{n,\psi}(f), \qquad Q_n(f) = \sum_{j=0}^{n} \rho_j f(x_j). \end{aligned} \tag{4.4}$$

Observe that

$$R_{n,\psi}(P_{2n}) = 0, \tag{4.5}$$

where $P_s(t)$ is an arbitrary polynomial, of degree $\leq s$, and that $Q_n(f)$ is independent of the choice of $\psi(t)$, even if the moment problem (2.1) is indeterminate.

We wish to study the conditions on $f(t)$ and $\psi(t)$ under which (4.4) converges, that is,

$$\lim_n R_{n,\psi}(f) = 0, \tag{4.6}$$

or, more generally,

$$\lim_{\nu} R_{n_\nu,\psi}(f) = 0, \tag{4.7}$$

for a certain infinite sequence of positive integers

$$n_1 < n_2 < \cdots < n_\nu < \cdots; \qquad n_\nu \to \infty.$$

Whenever we wish to cover simultaneously both the Stieltjes and Hamburger cases we write (4.4) as

$$I_\psi(f) \equiv \int_a^\infty f(t)\, d\psi = Q_n(f) + R_{n,\psi}(f), \tag{4.8}$$

where $a = 0$ or $-\infty$. In the Stieltjes case it is assumed that all roots of $q_{n+1}(z)$ lie in $[0, \infty)$.

We shall write $Q_n^{[c,d]}$, $Q_n'^{[c,d]}$, in case the summation is extended over all $x_{jn}$ in $[c, d]$ or outside $[c, d]$ respectively.

Taking

$$q_{n+1}(z) = \omega_{n+1}(z) - \frac{\omega_{n+1}(x_0)}{\omega_n(x_0)}\omega_n(z), \tag{4.9}$$

where $x_0$ is real and is not a root of $\omega_n(z)$, we have an approximate quadrature formula with a preassigned abscissa $x_0$.

The following two particular cases of (4.4) deserve special consideration.

Case A. $$q_{n+1}(z) = \omega_{n+1}(z).$$

Case C. $$q_{n+1}(z) = \omega_{n+1}(z) - \frac{\omega_{n+1}(0)}{\omega_n(0)}\omega_n(z).$$

These polynomials are, up to constant factors, the denominators, $N_{2n+2}(z)$, $N_{2n+1}(z)$, of the even or odd approximants, respectively, of the corresponding continued fraction $\mathcal{C}(z)$, whose numerators will be denoted by $Z_n(z)$. The polynomials $\omega_{n+1}(z) \equiv \omega_{n+1}(z;\, a,\, \infty;\, d\psi)$ are also proportional to the denominators of the successive approximants of the associated continued fraction $\mathfrak{A}(z)$. In the Stieltjes case the polynomials of case C always exist since $\omega_n(0) \neq 0$, $n = 1, 2, \cdots$. In the Hamburger case $\mathcal{C}(z)$ may not exist. Then we choose $x = \lambda$ such that $\omega_n(\lambda) \neq 0$, $n = 1, 2, \cdots$. By the substitution $x' = x - \lambda$ we get a transformed moment problem, with moments $\{\mu_n'\}$, such that $\sum_0^\infty \mu_n' z^{-n-1}$ admits both the associated and the corresponding continued fractions. Thus we may say that in both the Stieltjes and Hamburger cases we have two approximate quadrature formulas based respectively upon the roots of the denominators of the even $\left\{\frac{Z_{2n}(z)}{N_{2n}(z)}\right\}$ or odd $\left\{\frac{Z_{2n-1}(z)}{N_{2n-1}(z)}\right\}$ approximants of the corresponding continued fraction $\mathcal{C}(z)$. We use the following notation.

$$I_\psi(f) = \sum_{j=0}^{n} A_{jn} f(\xi_{jn}) + R_{n,\psi}^A(f) \equiv Q_n^A(f) + R_{n,\psi}^A(f), \tag{4.10}$$

where $\xi_j \equiv \xi_{jn}$, $j = 0, 1, \cdots, n$, are the roots of $\omega_{n+1}(z)$, and

$$A_{jn} = \frac{Z_{2n+2}(\xi_j)}{N'_{2n+2}(\xi_j)} = \frac{1}{K_{n+1}(\xi_j)} = \frac{1}{K_n(\xi_j)} = \rho_n(\xi_j). \tag{4.11}$$

$$I_\psi(f) = \sum_{j=0}^{n} C_{jn} f(\eta_{jn}) + R^C_{n,\psi}(f) \equiv Q^C_n(f) + R^C_{n,\psi}(f), \tag{4.12}$$

where $\eta_j \equiv \eta_{jn}$, $j = 0, 1, \cdots, n$, are the roots of $N_{2n+1}(z)$, and

$$C_{jn} = \frac{Z_{2n+1}(\eta_j)}{N'_{2n+1}(\eta_j)}. \tag{4.13}$$

Observe that

$$R^A_{n,\psi}(P_{2n+1}) = R^C_{n,\psi}(P_{2n}) = 0. \tag{4.14}$$

We shall have occasion to refer to the moment problem

$$\int_0^\infty t^n \, d\psi = \mu_{n+1}, \qquad n = 0, 1, \cdots. \tag{4.15}$$

Here we use the superscript 1: $d\psi^1(t)$, $A^1_{jn}$, $\omega^1_n(z)$, $\cdots$. Since

$$\int_a^\infty N_{2n+1}(t) P_{n-1}(t) \, d\psi = 0,$$

we have

$$N_{2n+1}(z) = \text{constant} \cdot z\omega^1_n(z). \tag{4.16}$$

Consider, first, the Stieltjes case. Here $d\psi^1(t) = t\, d\psi(t)$ yields a distribution function, so that

$$0 = \eta_{0n} < \eta_{1n} < \cdots < \eta_{nn},$$

whence, making use of (4.3),

$$C_{jn} = \frac{A^1_{j-1,n-1}}{\eta_{jn}} > 0, \qquad \eta_{jn} = \xi^1_{j-1,n-1}, \qquad j = 1, 2, \cdots, n. \tag{4.17}$$

In the Hamburger case $\psi^1(t)$, as introduced above, is not constantly increasing in $(-\infty, \infty)$. Formulas (4.17) still hold, but $A^1_{j-1,n-1}$ has the sign of $\eta_{jn}$, $j = 1, 2, \cdots, n$, and is not necessarily positive.

In view of (4.3) and of (II,**8**), we have

$$Q^A_n\left(\frac{1}{z-t}\right) = \frac{Z_{2n+2}(z)}{N_{2n+2}(z)}, \qquad Q^C_n\left(\frac{1}{z-t}\right) = \frac{Z_{2n+1}(z)}{N_{2n+1}(z)}. \tag{4.18}$$

**2.** The following results are readily established for the coefficients and abscissas of the approximate quadrature formulas.

(i) A linear transformation of the orthogonality interval $(a, b)$ does not change the coefficients $A_{jn}$.

(ii) In the "symmetric case", $\mu_1 = \mu_3 = \mu_5 = \cdots 0$, [Shohat, 5]

$$(4.19) \qquad \xi_{jn} = -\xi_{n-j,n}\,; \qquad A_{j,n} = A_{n-j,n}\,; \qquad j = 0, 1, \cdots, n.$$

$$(4.20) \quad \begin{cases} \omega_{2n}(z; -\infty, \infty; d\psi) = \omega_n(z^2; 0, \infty; d\psi_1), & d\psi_1(t) = 2\,d\psi(\sqrt{t}), \quad t \geqq 0, \\ \omega_{2n+1}(z; -\infty, \infty; d\psi) = z\omega_n(z^2; 0, \infty; d\psi_2), & d\psi_2(t) = 2t\,d\psi(\sqrt{t}), \\ n = 0, 1, 2, \cdots. \end{cases}$$

It follows, with obvious notations, that

$$(4.21) \quad \begin{cases} A_{j,2n-1}(-\infty, \infty; d\psi) = \tfrac{1}{2}A_{j,n-1}(0, \infty; d\psi_1) & = 0, 1, \cdots, n-1. \\ A_{j,2n}(-\infty, \infty; d\psi) = \dfrac{A_{j,n-1}(0, \infty; d\psi_2)}{2\xi_{j,n-1}(0, \infty; d\psi_2)} & \\ A_{n,2n}(-\infty, \infty; d\psi) = \dfrac{1}{K_n(0; -\infty, \infty : d\psi)} = \dfrac{1}{K_n(0; 0, \infty; d\psi_1)}. & \end{cases}$$

(iii) The interval $[x_{jn}, x_{j+1,n}]$ formed by any two consecutive roots of $q_{n+1}(z)$ contains at least one root of $q_{n+k}(z)$, $k > 1$. In particular, between any two consecutive roots of $\omega_{n+1}(z)$ lies at least one root of $\omega_{n+k}(z)$, $k > 1$ [cf. II,7, III,**2**].

Indeed, if $[x_{jn}, x_{j+1,n}]$ contains no roots of $q_{n+k}(z)$, then it is an interval of constancy for $\psi_{n+k}(t)$, and, by Corollary 2.2, $x_{jn}$ and $x_{j+1,n}$ could not have been roots of $q_{n+1}(z)$.

In (II,**7**) it was shown that

$$(4.22) \quad a < \xi_{0,n} < \xi_{0,n-1} < \xi_{1,n} < \xi_{1,n-1} < \cdots < \xi_{n-1,n-1} < \xi_{n,n} < b.$$

By a similar reasoning, it is shown that in the Stieltjes case

$$(4.23) \qquad 0 < \xi_{0,n} < \xi^1_{0,n} < \xi_{1,n} < \xi^1_{1,n} < \cdots < \xi_{n,n} < \xi^1_{n,n}.$$

Finally observe that for the polynomial (4.2)

$$(4.24) \quad \begin{array}{ll} x_{0n} < \xi_{0n} < x_{1n} < \cdots < x_{nn} < \xi_{nn}, & \text{if} \quad A_n > 0, \\ \xi_{0n} < x_{0n} < \xi_{1n} < \cdots < \xi_{nn} < x_{nn}, & \text{if} \quad A_n < 0. \end{array}$$

(iv) The following properties of the coefficients $A_{jn}$ are readily derived from (4.3) upon observing that $K'_{n+1}(x) = 2\sum_0^{n+1} \omega_i'(x)\omega_i(x)$ is positive (negative) for $x > \xi_{n,n}(x < \xi_{0,n})$:

$$(4.25) \quad \begin{array}{l} A_{0,n} > A_{0,n+1} > A_{0,n+2} > \cdots > \dfrac{1}{K_n(a)} = \rho_n(a) \\ A_{n,n} > A_{n+1,n+1} > A_{n+2,n+2} > \cdots > \dfrac{1}{K_n(b)} = \rho_n(b). \end{array}$$

Hence,

$$\lim_n A_{0n} \geqq \rho(a), \qquad \lim_n A_{nn} \geqq \rho(b).$$

**3.** We now turn to the study of the convergence properties of the approximate quadrature formulas. With the notation of Chapter II, we introduce the function $\psi_n(t) \equiv \psi_n^Q(t)$, special cases of which are $\psi_n^A(t)$, $\psi_n^C(t)$ in (4.10) and (4.12) respectively. We have

$$Q_n(f) = \int_a^\infty f(t)\,d\psi_n^Q \equiv I_{\psi_n}^Q(f), \tag{4.26}$$

$$I_{\psi_n^A}(t^k) = \mu_k, \qquad k = 0, 1, \cdots, 2n+1, \tag{4.27}$$

$$I_{\psi_n^C}(t^k) = \mu_k, \qquad k = 0, 1, \cdots, 2n. \tag{4.28}$$

$$\begin{cases} \dfrac{p_n(z)}{q_n(z)} = I_{\psi_n}\left(\dfrac{1}{z-t}\right); \quad \dfrac{Z_{2n+2}(z)}{N_{2n+2}(z)} = I_{\psi_n^A}\left(\dfrac{1}{z-t}\right) \\ \qquad \dfrac{Z_{2n+1}(z)}{N_{2n+1}(z)} = I_{\psi_n^C}\left(\dfrac{1}{z-t}\right). \end{cases} \tag{4.29}$$

We see that in (4.8), (4.10) or (4.12) we may take for $\psi(t)$ any solution of the reduced moment problem

$$\int_a^\infty t^k\,d\psi = \mu_k, \qquad k = 0, 1, \cdots, 2n \quad \text{or} \quad 2n+1 \text{ respectively.} \tag{4.30}$$

We see further that the convergence of the approximate quadrature formulas is equivalent to

$$\lim_\nu I_{\psi_{n_\nu}}(f) = I_\psi(f). \tag{4.31}$$

Limit-functions of convergent subsequences $\{\psi_{n_\nu}^Q(t)\}$, $\{\psi_{n_\nu}^A(t)\}$, $\{\psi_{n_\nu}^C(t)\}$ will be denoted by $\psi(t) \equiv \psi^Q(t)$, $\psi^A(t)$, $\psi^C(t)$ respectively The following theorem is now readily established

THEOREM 4.1. *The following three statements are equivalent:*

α) $\psi_{n_\nu}(t)$ *converges substantially;*

β) $Q_{n_\nu}\left(\dfrac{1}{z-t}\right)$ *converges for $z$ not on* $[a, \infty]$;

γ) $\dfrac{p_{n_\nu}(z)}{q_{n_\nu}(z)}$ *converges for $z$ not on* $[a, \infty]$.

**4.** To extend convergence from the special function $f(t) = \dfrac{1}{z-t}$ to a wider class of functions we use M. Riesz' approximation theorem (cf. II,**19**), combined with the following lemmas which are direct consequences of the positive linear character of the operator $Q_n(f)$ and of the relation $\sum_{j=0}^n \rho_{jn} = \mu_0$.

For brevity we shall write $f \in F_{n_\nu}^Q$, $F_{n_\nu}^A$, $F_{n_\nu}^C$, if, respectively, $Q_{n_\nu}(f)$, $Q_{n_\nu}^A(f)$,

$Q^C_{n_\nu}(f)$ converges. If no emphasis is laid on $Q, A, C$ we write simply $f \in F_{n_\nu}$ or even $f \in F$.

The following results are obvious.

(i) $f \in F$ and $\varphi \in F$ implies $\alpha f + \beta\varphi \in F$, where $\alpha$ and $\beta$ are constants.

(ii) $|f_1(t) - f_2(t)| < \epsilon$ in $[a, \infty)$ implies $|R_{n,\psi}(f_1 - f_2)| < 2\mu_0\epsilon$.

(iii) If $f(t)$ can be approximated on $[a, \infty)$ arbitrarily closely and uniformly by a linear aggregate of functions of a certain type $g(t)$ and if each $g(t) \in F$, then $f(t) \in F$.

We shall say that $f(t)$ is continuous in $[a, \infty)$ if it is continuous in every closed finite subinterval and tends to a finite limit as $t$ becomes infinite. Applying M. Riesz' approximation theorem and using Theorem 4.1, we get

LEMMA 4.1. *If $f(t)$ is continuous in $[a, \infty)$, then:* (i) *there exist convergent subsequences $\{Q_{n_\nu}(f)\}$ of the approximate quadrature formula* (4.8); (ii) *$Q_{n_\nu}(f)$ converges for all $f$ if and only if $\left\{\dfrac{p_{n_\nu}(z)}{q_{n_\nu}(z)}\right\}$ converges ($z$ not on $[a, \infty)$) or, which is the same, if and only if $\psi^Q_{n_\nu}(t)$ converges; if $\psi^Q_{n_\nu}(t) \to \psi^Q(t)$, then $Q_{n_\nu}(f) \to I_{\psi^Q}(f)$.*

LEMMA 4.2. *If $\int_c^d f(t)\, d\psi^Q$ exists, where $c, d$ are two finite points of continuity of $\psi^Q(t)$, if $\varphi(t) = f(t)$ in $[c, d]$ and $= 0$ elsewhere, then $\varphi(t) \in F^Q_{n_\nu}$.*

COROLLARY 4.1. (i) $\lim\limits_{\nu} Q^{[c,d]}_{n_\nu}(f) = \int_c^d f(t)\, d\psi^Q$ *if $f \in F^Q_n$*; (ii) $\lim\limits_{\nu} Q^{[c,d]}_{n_\nu}(t^m) = \int_c^d t^m\, d\psi^Q$, $\lim\limits_{\nu} Q'^{[c,d]}_{n_\nu}(t^m) = \int_a^c t^m\, d\psi^Q + \int_d^\infty t^m\, d\psi^Q$, *where $m$ is any positive integer or zero. In particular,*

$$\lim_{\nu} \sum\nolimits^{[c,d]} \rho_{jn_\nu} = \int_c^d d\psi^Q; \qquad \lim_{\nu} \sum\nolimits'^{[c,d]} \rho_{jn_\nu} = \int_a^c d\psi^Q + \int_d^\infty d\psi^Q.$$

From Lemma 4.2 we derive in the same manner as in case of a finite interval [Shohat, 8]

LEMMA 4.3. *Any interval $[c, d]$ which is not an interval of constancy of $\psi^Q(t)$ contains at least one $z_{jn_\nu}$, for $\nu$ sufficiently large. In other words, if infinitely many $q_{n_\nu}(z)$ have no roots in $[c, d]$, then $\psi^Q(t)$ is constant in $[c, d]$. Moreover, if $\int_\alpha^\beta d\psi^Q > 0$ for every $[\alpha, \beta] \subset [c, d]$ then the roots of $q_{n_\nu}(z)$ are dense in $[c, d]$. In any such $[c, d]$ the distance between the consecutive roots $z_{jn_\nu}$, $z_{j+1,n_\nu}$, and also between the end-point and the nearest root, tends to zero, as $\nu \to \infty$.*

It follows that *every point of the spectrum of $\psi^Q(t)$ is a limit point of the roots of $\{q_{n_\nu}(z)\}$.*

LEMMA 4.4. *To any $\epsilon > 0$ there corresponds a $G_0 = G_0(\epsilon) > 0$ with the following property: for any fixed $G_1 \geqq G_0$ such that $\pm G_1$ are points of continuity of $\psi^Q(t)$, a positive integer $\nu_0 = \nu_0(G_1, \epsilon)$ may be chosen so that*

$$|I_{\psi^Q}(f) - Q^{[-G_1, G_1]}_{n_\nu}(f)| < \epsilon \quad \textit{for} \quad \nu \geqq \nu_0.$$

First, choose $G_0 = G_0(\epsilon)$ so that $\left| \int_{-\infty}^{-G} f(t)\, d\psi^Q \right| + \left| \int_{G_1}^{\infty} f(t)\, d\psi^Q \right| < \frac{\epsilon}{2}$, for any $G_1 \geqq G_0$. Now fix $G_1 \geqq G_0$ so that $\pm G_1$ are points of continuity of $\psi^Q(t)$, and take $f_1(t) = f(t)$ in $[-G_1, G_1]$ and $=0$ elsewhere. By Lemma 4.2, $\nu_0 = \nu_0(G_1, \epsilon)$ may be found such that

$$| I_{\psi^Q}(f_1) - Q_{n_\nu}(f_1) | = \left| \int_{-G_1}^{G_1} f(t)\, d\psi - Q_{n_\nu}^{[-G_1, G_1]}(f) \right| < \frac{\epsilon}{2}, \qquad \nu \geqq \nu_0.$$

COROLLARY 4.2. *A necessary and sufficient condition for the convergence of* $Q_{n_\nu}(f)$ *to* $I_{\psi^Q}(f)$ *is that to any given* $\epsilon > 0$ *there correspond sufficiently large* $G_1$ *and* $\nu_0$ *such that* $\psi^Q(t)$ *is continuous at* $\pm G_1$, *and* $| Q_{n_\nu}'^{[-G_1, G_1]}(f) | < \epsilon$ *for* $\nu \geqq \nu_0$.

COROLLARY 4.3. (i) *If* $\omega(t) \geqq 0$ *for* $|t| \geqq t_0$ *and if* $f(t) = O(\omega(t))$ *as* $|t| \to \infty$, *then* $\omega(t) \in F$ *implies* $f(t) \in F$.

(ii) *If* $f(t) = f_1(t) f_2(t)$, *then* $f(t) \in F_{n_\nu}$, *provided* $|f_1(t)|^p \in F_{n_\nu}$ *and* $Q_{n_\nu}'^{[-t_0, t_0]}(|f_2|^{p'}) = O(1)$, *as* $\nu \to \infty$, $\frac{1}{p} + \frac{1}{p'} = 1$.

(iii) $f(t) \in F$ *is implied by any one of the following conditions:* $\alpha$) $|f(t)|^p \in F$, $p \geqq 1$; $\beta$) $f(t) = O(|t|^m)$, *where* $m > 0$ *is fixed;* $\gamma$) $Q'^{[-t_0, t_0]}(|f(t)|^p) = O(1)$, $p > 1$.

Statements similar to those in Corollaries 4.2 and 4.3 have been given, under more restrictive conditions, by Jouravsky, whose considerations have certain points in common with those developed here [Jouravsky, 1].

Another direct consequence is the following theorem frequently used in the subsequent discussion.

THEOREM 4.2. *If, for* $|t|$ *sufficiently large and for* $\nu \geqq \nu_0$, $f(t)$ *and* $R_{n_\nu, \psi^Q}(f)$ *keep the same constant sign (zero values not excluded), then* $f(t) \in F_{n_\nu}^Q$. *Analogous results hold in the Stieltjes case* [Shohat, 3].

Assume, without loss of generality,

$$(4.32) \qquad f(t) \geqq 0, \qquad R_{n_\nu, \psi^Q}(f) \geqq 0, \qquad \text{for} \quad |t| \geqq t_0, \qquad \nu \geqq \nu_0.$$

In Lemma 4.4 take $G_0 > t_0$. Then

$$Q_{n_\nu}'^{[-G_1, G_1]}(f) + R_{n_\nu, \psi^Q}(f) = | I_{\psi^Q}(f) - Q_{n_\nu}^{[-G_1, G_1]}(f) | < \epsilon, \qquad \nu \geqq \nu_0,$$

and the result follows since both terms on the left are positive.

COROLLARY 4.4. *Let* $f(t) \geqq 0$ *in* $[a, \infty)$. *If there exists a solution* $\psi'(t)$ *and a certain sequence* $\{n_\nu'\}$ *such that* $Q_{n_\nu'}(f) \to I_{\psi'}(f)$, *then for any* $\psi^Q(t) = \lim_\nu \psi_{n_\nu}^Q(t)$, $\{n_\nu\} \subset \{n_\nu'\}$, $I_{\psi^Q}(f) \leqq I_{\psi'}(f)$.

We have $R_{n_\nu, \psi^Q}(f) = I_{\psi^Q}(f) - Q_{n_\nu}(f) \to I_{\psi^Q}(f) - I_{\psi}'(f)$. If $I_{\psi^Q}(f) > I_{\psi'}(f)$, then $R_{n_\nu,\psi_0}(f) > 0$, $\nu \geqq \nu_0$, and $Q_{n_\nu}(f) \to I_{\psi_0}(f)$, which contradicts the hypothesis

$$\lim_\nu Q_{n_\nu}(f) = \lim_\nu Q_{n_\nu'}(f) = I_{\psi'}(f).$$

**5.** We proceed to study the behavior of the abscissas, as $n \to \infty$. The inequalities (4.22) show that

$$\lim_n \xi_{0,n} = a_1 \geqq a, \qquad \lim_n \xi_{n,n} = b_1 \leqq \infty. \tag{4.33}$$

This implies that $[a_1, b_1]$ is the "true interval" of orthogonality for the orthonormal polynomials $\{\omega_n(z)\}$. This means: *there exists a solution $\psi_0(t)$ of the moment problem* (2.1) *which is constant outside* $[a_1, b_1]$, *while no solution of the moment problem* $\int_c^d t^n\, d\psi = \mu_n$, $n = 0, 1, \cdots$, *exists if* $[c, d] \subset [a_1, b_1]$.

Observe that any $\psi^A(t)$ is a solution of the moment problem (2.1), with spectrum in $[a_1, b_1]$.

Since we are dealing with approximate quadratures pertaining to an infinite interval we assume from now on that one, at least, of the extreme roots of $\omega_n(z)$ increases indefinitely in absolute value. Thus, in the Stieltjes case $a_1 \geqq 0$, $b_1 = \infty$. Stieltjes [5] has shown that the number of $\xi_{jn}$ which tend to infinity, as $n \to \infty$, is either zero or infinite and the number of $\xi_{jn}$ which tends to $a_1$ is either one or infinite. It follows, by (4.23) and (4.24), that in the Stieltjes case

$$x_{nn} \to \infty, \qquad \xi'_{nn} \to \infty;$$

$$\xi_{in} \to a_1, \quad i = 0, 1, \quad \text{implies} \quad \xi'_{0,n} \to a_1 \quad \text{and} \quad x_{0,n} \to a_1, \quad \text{if} \quad A_n < 0.$$

$$\xi_{0n} \to a_1 \quad \text{and} \quad \xi_{1n} \to a' > a_1 \quad \text{implies} \quad \xi^1_{0n} \to a^1_1, \qquad 0 \leqq a_1 < a^1_1 \leqq a'.$$

The reasoning of Stieltjes may be extended to $(-\infty, \infty)$ and shows that the number of roots of $\omega_n(z)$ which $\to \pm\infty$, as $n \to \infty$, is zero or infinite. Thus, in the Hamburger case at least one of the quantities $|\xi_{0n}|$, $|\xi_{nn}|$ tends to infinity. Furthermore,

$$\xi_{nn} \to \infty \quad \text{implies} \quad x_{nn} \to \infty; \qquad \xi_{0n} \to -\infty \quad \text{implies} \quad x_{0n} \to -\infty.$$

Next we study the dependence of the roots of $q_{n+1}(z) = \omega_{n+1}(z) + \alpha\omega_n(z)$ on the real parameter $\alpha$, for a given $n$. Let $\xi$ be such a root. By Darboux' formula [Shohat, 5]

$$K_n(z) = \frac{a_n}{a_{n+1}} [\omega'_{n+1}(z)\omega_n(z) - \omega'_n(z)\omega_{n+1}(z)],$$

we have

$$\frac{d\xi}{d\alpha} = -\frac{a_n}{a_{n+1}} \cdot \frac{\omega_n^2(\xi)}{K_n(\xi)} < 0. \tag{4.34}$$

It follows that *the roots $x_{jn}$ are decreasing functions of $\alpha$.* More precisely,

$$x_{0,n} = -\infty; \quad x_{jn} = \xi_{j-1,n-1}, \quad j = 1, 2, \cdots, n, \quad \text{if } \alpha = +\infty;$$

$$x_{0,n} < a, \quad \xi_{j-1,n-1} < x_{jn} < \xi_{jn}, \quad j = 1, 2, \cdots, n, \quad \text{if } \alpha > -\frac{\omega_{n+1}(a)}{\omega_n(a)};$$

$$\xi_{jn} < x_{jn} < \xi_{j,n-1}, \quad j = 0, 1, \cdots, n-1, x_{nn} > b, \quad \text{if } \alpha < -\frac{\omega_{n+1}(b)}{\omega_n(b)};$$

$$x_{jn} = \xi_{j,n-1}, \quad j = 0, 1, \cdots, n-1, \quad x_{nn} = +\infty, \quad \text{if } \alpha = -\infty.$$

$\alpha = -\dfrac{\omega_{n+1}(a)}{\omega_n(a)}$ or $-\dfrac{\omega_{n+1}(b)}{\omega_n(b)}$ implies $x_{0n} = a$ or $x_{nn} = b$ respectively.

Thus, a necessary and sufficient condition for all roots of $q_{n+1}(z)$ to lie in the orthogonality interval $[a, b]$ (finite) is: $-\dfrac{\omega_{n+1}(a)}{\omega_n(a)} \geqq \alpha \geqq -\dfrac{\omega_{n+1}(b)}{\omega_n(b)}$. In the Stieltjes case all roots of $q_{n+1}(z)$ lie in $[0, \infty)$ if and only if $-\dfrac{\omega_{n+1}(0)}{\omega_n(0)} \geqq \alpha$. It also follows that $\rho_n(x_{0n}) = \dfrac{1}{K_n(x_{0n})}$ *decreases and* $\rho_n(x_{nn}) = \dfrac{1}{K_n(x_{nn})}$ *increases, as $\alpha$ increases.*

The effect of the variation of the moments $\{\mu_n\}$ on the roots $\{\xi_{jn}\}$ has been studied by Markoff in the Stieltjes case. His result is [Markoff, 7]:

*If the even moments $\mu_0, \mu_2, \cdots, \mu_{2n}$ do not increase and the odd moments $\mu_1, \mu_3, \cdots, \mu_{2n+1}$ do not decrease, then $\Delta_0, \Delta_1, \cdots, \Delta_n$ do not increase and $\Delta_0^{(1)}, \Delta_1^{(1)}, \cdots, \Delta_n^{(1)}$ do not decrease as long as they all remain positive. Moreover, each root $\xi_{jn}$, $j = 0, 1, \cdots, n$, increases. It follows that the addition of a mass at $t = 0$ increases all $\Delta_n$ and decreases all $\Delta_n^{(1)}$; also each $\xi_{jn}$ decreases.*

**6.** We now turn to the coefficients.

THEOREM 4.3. *Let c, d be points of continuity of $\psi^Q(t)$. Denote by $\{x'_{n_\nu}\}$ the subsequence of the roots $\{x_{j,n_\nu}\}$ which, for $\nu \geqq \nu_0$, remain in $[c, d]$, by $\{\rho'_{n_\nu}\}$ the subsequence of the corresponding coefficients $\{\rho_{j,n_\nu}\}$. Then either the largest of the $\rho'_{n_\nu}$, say, $\rho^*_{n_\nu}$, tends to 0 as $\nu \to \infty$, or else, if $\lim_\nu \rho^*_{n_\nu} = \rho > 0$, then there exists at least one point $\alpha$ in $[c, d]$ where $\psi(\alpha + 0) - \psi(\alpha - 0) \geqq \rho$.*

The proof follows that of [Fejer, 1] and may be omitted.

COROLLARY 4.5. *If $\psi^Q(t)$ is continuous in $[a, \infty)$, then $\lim_\nu \rho_{j,n_\nu} = 0$ uniformly for $j = 0, 1, \cdots, n_\nu$.*

By 4.5,

$$I_{\psi^Q}(x^k) = \mu_k, \quad k = 0, 1, \cdots, 2n,$$

whence

$$\rho_{j,n_\nu} x^2_{j,n_\nu} < \mu_2, \quad j = 0, 1, \cdots, n_\nu. \tag{4.35}$$

Choose $G > 0$ so large that $\frac{\mu_2}{G^2} < \epsilon$. Our statement is proved by (4.35) and Theorem 4.3, $\psi(t)$ being continuous in $[-G, G]$. The continuity of $\psi(t)$ is essential as was shown in the case of a finite interval [Shohat, 8].

Much light is thrown on the behavior of the $\{\rho_{jn}\}$ and of a solution $\psi(t)$ by Tchebycheff's inequalities [Lemma 2.13]. We write them in full, denoting by $\psi(t)$ any solution of the moment problem (2.1) or of the reduced moment problem (2.9) and by $[a, b]$ the interval of orthogonality.

$$(4.36)\quad \begin{cases} 1. \quad \rho_{0n} + \rho_{1,n} + \cdots + \rho_{j-1,n} \\ \qquad \leqq \int^{x_{jn}-0} d\psi \leqq \int_a^{x_{jn}+0} d\psi \leqq \rho_{0n} + \rho_{1n} + \cdots + \rho_{jn}, \quad j = 1, 2, \cdots, n. \\ 2. \quad \int_{x_{j+1,n}+0}^{x_{k+1,n}-0} d\psi \geqq \rho_{jn} + \rho_{j+1,n} + \cdots + \rho_{kn} \geqq \int_{x_{jn}-0}^{x_{kn}+0} d\psi, \qquad k > j. \\ 3. \quad \int_a^{x_{0n}+0} d\psi \leqq \rho_{0n}; \qquad \int_{x_{nn}-0}^{b} d\psi \leqq \rho_{nn}. \\ 4. \quad \int_{x_{j-1,n}+0}^{x_{j+1,n}-0} d\psi > \rho_{jn} \geqq \int_{x_{jn}-0}^{x_{jn}+0} d\psi, \\ \qquad\qquad j = 0, 1, \cdots, n, \qquad x_{-1,n} \equiv a, \qquad x_{n+1,n} \equiv b. \\ 5. \quad \int_{x_{j-1,n}+0}^{x_{j+2,n}-0} d\psi \geqq \rho_{jn} + \rho_{j+1,n} \geqq \int_{x_{jn}-0}^{x_{j+1,n}+0} d\psi. \\ 6. \quad \psi(x_{jn} - 0) \leqq \sigma_{jn} \equiv \rho_{1n} + \rho_{2n} + \cdots + \rho_{jn} \leqq \psi(x_{j+1,n} + 0), \\ \qquad\qquad j = 0, 1, \cdots, n-1. \\ 7. \quad \sigma_{jn} = \psi(\theta_{jn}), \qquad x_{jn} \leqq \theta_{jn} \leqq x_{j+1,n}; \qquad \rho_{jn} = \psi(\theta_{jn}) - \psi(\theta_{j-1,n}), \\ \qquad\qquad \text{if } \psi(t) \text{ is continuous in } (a, b). \\ 8. \quad \rho_{0n} + \rho_{1n} + \cdots + \rho_{j-2,n} \leqq \int_a^{\alpha} d\psi \leqq \rho_{0n} + \rho_{1n} + \cdots + \rho_{jn}, \\ \qquad\qquad \text{if } x_{j-1,n} \leqq \alpha \leqq x_{jn}. \\ 9. \quad \rho_{i+1,n} + \rho_{i+2,n} + \cdots + \rho_{j-2,n} \leqq \int_{\alpha}^{\beta} d\psi \leqq \rho_{i-1,n} + \rho_{in} + \cdots + \rho_{jn}, \\ \qquad\qquad \text{if } x_{i-1,n} \leqq \alpha \leqq x_{in} \text{ and } x_{j-1,n} \leqq \beta \leqq x_{jn}, \quad j > i. \end{cases}$$

These inequalities, combined with the preceding statements, yield the following results.

(i) Let $\psi(t)$ be continuous in the finite interval $[c, d] \subset [a, b]$. Denote by $x_{k,n_\nu}, x_{k+1,n_\nu}, \cdots, x_{l,n_\nu}$ the roots of $q_{n_\nu+1}(z)$ which are in $[c, d]$, for $\nu \geqq \nu_0$ ($k$, $l$ eventually depending on $n$). Then $\lim_\nu \int_{x_{i,n_\nu}}^{x_{i+1,n_\nu}} d\psi = 0$, uniformly for

$i = k, k+1, \cdots, l-1$. In particular, if $\psi^Q(t)$ is continuous in $[a, b]$, this limiting relation holds uniformly for $i = 0, 1, \cdots, n-1$.

(ii) Let $\psi^Q(t)$ be continuous in the finite subinterval $[c, d]$ which contains no interval of constancy for $\psi^Q(t)$. If, for any infinite subsequence $\{n_\nu\}$ and for a fixed $i$, $x_{i-1,n_\nu}$, $x_{i+1,n_\nu}$ remain in $[c, d]$ for $\nu \geqq \nu_0$, then $\lim_\nu \rho_{i,n_\nu} = 0$.

We may obtain various other inequalities for the coefficients by specifying $\psi(t)$ in (4.36). Thus, for $\psi(t) = \psi^A_{n+1}(t)$, we get

$$A_{0,n+1} < A_{0n} < A_{0,n+1} + A_{1,n+1} < A_{0n} + A_{1n} < \cdots < A_{0,n+1} + \cdots + A_{n,n+1}$$
$$< A_{0,n} + \cdots + A_{nn} = A_{0,n+1} + \cdots + A_{n+1,n+1}.$$

More generally, take in (4.36) $\psi(t) = \psi'_n(t), \psi''_n(t)$, corresponding respectively to $q'_{n+1}(z) = \omega_{n+1}(z) + \alpha'\omega_n(z)$ and $q''_{n+1}(z) = \omega_{n+1}(z) + \alpha''\omega_n(z)$. In view of the dependence of $\{x_{jn}\}$ on $\alpha$, as stated above, we get

$$\alpha'' > \alpha' \quad \text{implies} \quad \rho''_{0n} < \rho'_{0n} < \rho''_{0n} + \rho''_{1n} < \rho'_{0n} + \rho'_{1n} < \cdots$$
$$< \rho'_{0n} + \cdots + \rho'_{n-1,n}, \qquad \rho'_{nn} < \rho''_{nn}.$$

In particular, taking $\alpha'' = 0$, or $\alpha' = 0$ and $\alpha'' = -\dfrac{\omega_{n+1}(0)}{\omega_n(0)}$, we have

$$\alpha' < 0 \quad \text{implies} \quad A_{0n} < \rho'_{0n} < A_{0n} + A_{1n} < \rho'_{0n} + \rho'_{1n} < \cdots$$
$$< \rho'_{0n} + \cdots + \rho'_{n-1,n} < A_{0n} + \cdots + A_{nn}, \qquad \rho'_{nn} < A_{nn};$$

$$\alpha' > 0 \quad \text{implies} \quad \rho'_{0n} < A_{0n} < \rho'_{0n} + \rho'_{1n} < A_{0n} + A_{1n} < \cdots$$
$$< A_{0n} + \cdots + A_{n-1,n}, \qquad A_{nn} < \rho'_{nn};$$

$$\frac{\omega_{n+1}(0)}{\omega_n(0)} < 0 \quad \text{implies} \quad C_{0n} = \frac{1}{K_n(0)} < A_{0n} < C_{0n} + C_{1n}$$
$$< A_{0n} + A_{1n} < \cdots < A_{0n} + \cdots + A_{n-1,n}, \qquad A_{nn} < C_{nn};$$

$$\frac{\omega_{n+1}(0)}{\omega_n(0)} > 0 \quad \text{implies} \quad A_{0n} < C_{0n} = \frac{1}{K_n(0)} < A_{0n} + A_{1n} < C_{0n} + C_{1n}$$
$$< \cdots < C_{0n} + \cdots + C_{n-1,n}, \qquad C_{nn} > A_{nn}.$$

Thus, in the Stieltjes case, by (4.17),

$$\frac{1}{K_n(0)} < A_{0n} < \frac{1}{K_n(0)} + \frac{A^1_{0,n-1}}{\xi^1_{0,n-1}} < A_{0n} + A_{1n}$$

$$(4.37) \qquad < \frac{1}{K_n(0)} + \frac{A^1_{0,n-1}}{\xi^1_{0,n-1}} + \frac{A^1_{1,n-1}}{\xi^1_{1,n-1}} < \cdots < A_{0,n} + \cdots + A_{n-1,n}$$

$$< \frac{1}{K_n(0)} + \frac{A^1_{0,n-1}}{\xi^1_{0,n-1}} + \cdots + \frac{A^1_{n-1,n-1}}{\xi^1_{n-1,n-1}} = \mu_0, \qquad A_{nn} < \frac{A^1_{n-1,n-1}}{\xi^1_{n-1,n-1}}.$$

We see that

$$\sum_{j=0}^{n} \frac{A_{j,n}^1}{\xi_{j,n}^1} = \mu_0 - \frac{1}{K_{n+1}(0)} \uparrow \mu_0 - \rho(0). \tag{4.38}$$

The right-hand member is the value of the associated continued fraction $\mathfrak{A}^1(z)$ at $z = 0$. We supplement (4.25) by

THEOREM 4.4. *Let* $\lim_n \xi_{0,n} = a_1$, $\lim_n \xi_{nn} = b_1$. *Then* $\lim_n A_{0n} = \rho(a_1)$ *and* $\lim_n A_{nn} = \rho(b_1)$.

It suffices to give a proof for $A_{0n}$. If $a_1 = -\infty$, the statement follows from the inequality

$$A_{0n} < \frac{\mu_2}{\xi_{0n}^2}.$$

If $a_1$ is finite, we use the inequalities

$$\xi_{0k} > \xi_{0n} > a_1, \quad \text{if} \quad k > n; \qquad \frac{1}{K_k(\xi_{0k})} > \frac{1}{K_k(\xi_{0n})} > \frac{1}{K_n(\xi_{0n})} = A_{0n},$$

$$\rho(a_1) = \lim_n \frac{1}{K_n(a_1)} < \frac{1}{K_n(a_1)} < \frac{1}{K_n(\xi_{0n})}.$$

The desired result follows if we choose $k$ so large that

$$0 < \frac{1}{K_k(a_1)} - \rho(a_1) < \frac{\epsilon}{2},$$

and then choose $n_0 > k$ so large that

$$0 < \frac{1}{K_k(\xi_{0n})} - \frac{1}{K_k(a_1)} < \frac{\epsilon}{2}, \qquad n \geqq n_0 > k.$$

**7.** We now turn to a discussion of $\psi(t)$ and of the associated and corresponding continued fractions. On putting $\lambda = 0$ in Corollary 2.5, we get

THEOREM 4.5. *The sequence of odd approximants to the corresponding continued fraction* $\mathcal{C}(z)$ *always converges (and yields the extremal solution of the moment problem* (2.1) *with maximal mass at* $t = 0$).

The reason for this fundamental result of Stieltjes and Hamburger [Stieltjes, 5; Hamburger, 3] lies in the fact that all $\psi_n^C(t)$ have a concentrated mass at the fixed point $t = 0$.

Regarding convergence of the even approximants of $\mathcal{C}(z)$ (i.e. of the associated continued fraction $\mathfrak{A}(z)$), nothing definite can be said in the general Hamburger case. In the Stieltjes case they also converge for $z$ not on $[0, \infty)$ [Stieltjes, 5]. This we shall prove later on the basis of approximate quadratures.

The convergence of $\mathfrak{A}^1(z)$, for $z = 0$, was previously stated. We give here a proof of convergence for $z$ not on $[0, \infty)$.

From the known relations

$$\frac{\chi_n(z)}{\omega_n(z)} = \frac{Z_{2n}(z)}{N_{2n}(z)},$$

$$N_{2n-1}(z) = Az\omega^1_{n-1}(z),$$

where $A$ is constant, and

$$Z_n(z) = \int_0^\infty \frac{N_n(z) - N_n(t)}{z - t}\, d\psi, \tag{4.39}$$

we get readily

$$\begin{aligned} A^{-1} Z_{2n-1}(z) &= \int_0^\infty \frac{\omega^1_{n-1}(z) - \omega^1_{n-1}(t)}{z - t}\, d\psi^1 + \mu_0 \omega^1_{n-1}(z) \\ \frac{Z_{2n-1}(z)}{N_{2n-1}(z)} &= \frac{1}{z}\frac{\chi^1_{n-1}(z)}{\omega^1_{n-1}(z)} + \frac{\mu_0}{z}. \end{aligned} \tag{4.40}$$

This relation connects the odd approximants of $\mathcal{C}(z)$ with the approximants of $\mathcal{A}^1(z)$. The convergence of $\mathcal{A}^1(z)$ now follows from Theorem 4.5. It also follows that

$$\begin{aligned} \int_0^\infty \frac{d\psi^C(t)}{z - t} &= \frac{1}{z}\int_0^\infty \frac{d\psi^{A^1}(t)}{z - t} + \frac{\mu_0}{z}, \quad z \text{ not on } [0, \infty), \\ I(z; \psi) - \frac{Z_{2n+1}(z)}{N_{2n+1}(z)} &= \frac{1}{z}\left[I(z; \psi^1) - \frac{Z^1_{2n}(z)}{N^1_{2n}(z)}\right]. \end{aligned} \tag{4.41}$$

**8.** We now return to the study of the approximate quadrature formulas. We need expressions for the remainders $R^A_{n,\psi}$, $R^C_{n,\psi}$. We assume that $f(t)$ has all derivatives needed, and use the relations

$$R^A_{n,\psi}(f) = R^A_{n,\psi}(f - P_{2n+1}); \qquad R^C_{n,\psi}(f) = R^C_{n,\psi}(f - P_{2n}). \tag{4.42}$$

Take the polynomials $P_{2n+1}(t)$ and $P_{2n}(t)$ so that

$$P_{2n+1}(\xi_{jn}) = f(\xi_{jn}), \qquad P'_{2n+1}(\xi_{jn}) = f'(\xi_{jn}), \qquad j = 0, 1, \cdots, n; \tag{4.43}$$

$$P_{2n}(0) = f(0), \quad P_{2n}(\eta_{jn}) = f(\eta_{jn}), \quad P'_{2n}(\eta_{jn}) = f'(\eta_{jn}), \quad j = 1, 2, \cdots, n. \tag{4.44}$$

The following observation is due to Markoff [3]. Applying (4.10) to $P_{2n+1}(t)$ in (4.43) (Hermite interpolation polynomial) we get an expression of the form

$$\int_a^\infty P_{2n+1}(t)\, d\psi = \sum_{j=0}^{n} E_j f(\xi_j) + \sum_{j=0}^{n} D_j f'(\xi_j),$$

where the constants $E_j$ and $D_j$ do not depend on $f(t)$. The explicit expression of $D_j$ shows at once that all $D_j$ vanish if and only if the $\xi_j$ are the roots of $\omega_{n+1}(z)$.

A similar remark applies to (4.12). Making use of (4.17), we write, with obvious notations,

$$Q^C_n(f) = \rho_n(0) f(0) + Q^{A^1}_{n-1}\left(\frac{f(t)}{t}\right), \tag{4.45}$$

and employ on the right an interpolation polynomial $P_{2n-1}(t)$ analogous to (4.43). Formula (4.43) leads to the following interpolation formula with remainder,

$$f(t) = P_{2n+1}(t) + \frac{\omega_{n+1}^2(t)}{(2n+2)!\,a_{n+1}^2} f^{(2n+2)}(\xi), \qquad \xi \text{ in } [a, \infty).$$

A similar formula can be written for (4.44). This gives at once desired remainders in the approximate quadrature formulas (4.10), (4.12), namely,

$$(4.46) \qquad R_{n,\psi}^A(f) = \frac{f^{(2n+2)}(\xi)}{(2n+2)!\,a_{n+1}^2},$$

$$(4.47) \qquad R_{n,\psi}^C(f) = \frac{f^{(2n+1)}(\xi)}{(2n+1)!\,(a_n^1)^2}, \qquad a = 0,$$

$$(4.48) \qquad R_{n,\psi}^C(f) = \frac{f^{*(2n)}(\xi)}{(2n)!\,(a_n^1)^2}, \qquad f^*(t) = \frac{f(t) - f(0)}{t}, \qquad a = 0.$$

We now state

LEMMA 4.5. *For any $f(t)$, the two approximate quadrature sums $Q_n^C(f)$ and $Q_n^{A^1}\left(\frac{f(t)}{t}\right)$ converge or diverge simultaneously. More precisely,*

$$\lim_n \left[ Q_n^C(f) - Q_{n-1}^{A^1}\left(\frac{f(t)}{t}\right)\right] = f(0)\rho(0).$$

The following relations are direct consequences of (4.46), (4.47):

$$R_{n,\psi}^A(P_{2n+2}) = \frac{c_{2n+2}}{a_{n+1}^2}; \qquad R_{n,\psi}^C(P_{2n+1}) = \frac{c_{2n+1}}{(a_n^1)^2}, \qquad a = 0,$$

where $c_s$ is the highest coefficient of $P_s(t)$;

$$\int_a^\infty t^{2s}\,d\psi - \int_a^\infty t^{2s}\,d\psi_n^A > 0, \qquad s \geq n+1;$$

$$\int_0^\infty t^s\,d\psi - \int_0^\infty t^s\,d\psi_n^A > 0, \qquad s \geq 2n+2;$$

$$\int_0^\infty t^s\,d\psi - \int_0^\infty t^s\,d\psi_n^C > 0, \qquad s \geq 2n+1.$$

These inequalities have been used by Uspensky [2] and Jouravsky [1] in the study of $Q_n^A(f)$.

DEFINITION 4.1. *A function $f(t)$ is said to belong to the class $(1, n_\nu)$ or $(2, n_\nu)$ in $[a, \infty]$ if $f(t)$ and, respectively, an infinite sequence $\{f^{(2n_\nu+1)}(t)\}$ of its odd derivatives or $\{f^{(2n_\nu+2)}(t)\}$ of its even derivatives keep the same and constant sign in $[t_0, \infty)$ (which we may take as positive, zero values not excluded), where $t_0 \geq 0$ is sufficiently large, if $a = 0$, and $t_0 = -\infty$, if $a = -\infty$.*

Combining (4.46), (4.47), (4.48), Theorem 4.2 and Lemma 4.5, we arrive at the following fundamental convergence theorem.

THEOREM 4.6. *Let* $f(t) = O(G(t))$ *as* $t$ *becomes infinite. Then* (i) $G(t) \in (2, n_\nu)$ *implies* $f(t)$, $|f(t)| \in F^A_{n_\nu}$; (ii) $G(t) \in (1, n_\nu)$ *implies* $f(t)$, $|f(t)| \in F^C_{n_\nu}$, $a = 0$; (iii) $\dfrac{G(t) - G(t_0)}{t - t_0} \in (2, n_\nu)$ *in* $[c, \infty)$ *implies* $f(t)$, $|f(t)| \in F^A_{n_\nu}$, $c \geqq a = 0$.

We give a few examples of functions which belong to $(1, n)$ or $(2, n)$.

$f(t) = e^{kt}$, $k > 0$, belongs to both $(1, n)$ and $(2, n)$ in $[0, \infty)$. The integral transcendental function $f(t) = \sum_{n=0}^{\infty} c_n t^{2n}$ belongs to $(2, n)$ in $(-\infty, \infty)$, if all $c_n \geqq 0$. If $f(t)$ is completely monotonic in $[0, \infty)$, then $f(t) \in (2, n)$, as is seen from the canonical representation

$$f(t) = \int_0^\infty e^{-tx}\, d\alpha(x), \qquad t \geqq 0,$$

where $\alpha(x)$ is bounded and non-decreasing in $[0, \infty)$.

$$f(t) = \frac{1}{t - c} \in (2, n) \quad \text{in} \quad [0, \infty), \qquad c < 0$$

$$f(t) = \frac{t}{t - c} \in (1, n) \quad \text{in} \quad [0, \infty), \qquad c < 0$$

$$f(t) = \frac{e^t}{(1 + t)^\rho} \in (2, n) \quad \text{in} \quad [0, \infty), \qquad \rho \geqq 0,$$

for, $\Gamma(\rho)f(t) = \int_0^\infty e^{-s+(1-s)t} s^{\rho-1}\, ds$.

More generally, if $f(t)$ is completely monotonic in $[c, \infty)$, $c \geqq 0$, then $f(t) \in (2, n)$ in $(c, \infty)$. Thus,

$$(4.49) \quad f(t) = \frac{e^t}{(1 + t)^\rho \log^{1+\rho_1}(1 + t) \log_2^{1+\rho_2}(1 + t) \cdots \log_m^{1+\rho_m}(1 + t)} \in (2, n)$$

in $[c, \infty]$, $c$ sufficiently large, $\rho \geqq 0$; $\rho_1, \rho_2, \cdots, \rho_m > 0$,

for $f(t)e^{-t}$ is completely monotonic in $(c, \infty)$ [Hausdorff, 1, I].

**9.** We now state the following theorem which supplements Theorem 4.2.

THEOREM 4.7. *The function* $f(t)$ *being unrestricted as to sign, assume that* $R_{n'_\nu;\psi}(f)$ *has a constant sign for all* $n'_\nu$ *and all* $\psi$. *If there exists a solution* $\psi'$ *such that* $Q_{n_\nu}(f) \to I_{\psi'}(f)$, $\{n_\nu\} \subset \{n'_\nu\}$, *then* $I_{\psi'}(f)$ *gives the minimum (or maximum) of* $I_\psi(f)$, *according as* $R_{n'_\nu;\psi}(f)$ *is* $\geqq 0$ *or* $\leqq 0$.

In fact,

$$0 \leqq R_{n_\nu;\psi}(f) = I_\psi(f) - Q_{n_\nu}(f) \to I_\psi(f) - I_{\psi'}(f).$$

We state a further result where it is assumed that $f(t) \geqq 0$.

LEMMA 4.6. *If there exists a sequence* $\{n'_\nu\}$ *such that* $R_{n'_\nu;\psi^Q}(f) \geqq 0$ *for* $\psi^Q = \lim_\nu \psi^Q_{n_\nu}$, $\{n_\nu\} \subset \{n'_\nu\}$, *then* $I_{\psi^Q}(f)$ *has the same value for all such* $\psi^Q$. *Moreover,* $Q_{n'_\nu}(f) \to I_{\psi^Q}(f)$.

Extract from $\{\psi_{n_i}\}$ any two convergent subsequences $\{\psi'_\nu\}$, $\{\psi''_\nu\}$, with limit functions $\psi'$, $\psi''$ respectively. By theorem 4.2, $Q_{\nu'}(f) \to I_{\psi'}(f)$, $Q_{\nu''}(f) \to I_{\psi''}(f)$. Moreover,

$$R_{\nu',\psi''}(f) \to I_{\psi''}(f) - \lim_{\nu'} Q_{\nu'}(f) = I_{\psi''}(f) - I_{\psi'}(f) \geqq 0.$$

$$R_{\nu'',\psi'}(f) \to I_{\psi'}(f) - I_{\psi''}(f) \geqq 0.$$

The preceding general considerations may be applied to functions of classes $(1, n_\nu)$, $(2, n_\nu)$, for which the remainder in the approximate quadrature formulas (4.10), (4.12) keeps a constant sign.

THEOREM 4.8. *If* (i) $f(t) \in (2, n'_\nu)$ *in* $[a, \infty)$ *or* (ii) *either* $f(t) \in (1, n'_\nu)$ *or* $\dfrac{f(t)-f(0)}{t}$ $\in (2, n'_\nu)$ *in* $[0, \infty)$, *then* $I_{\psi^A}(f)$ *has the same value for any* $\psi^A = \lim_\nu \psi^A_{n_\nu}$, $\{n_\nu\} \subset \{n'_\nu\}$ *in* (i), *or* $I_{\psi^C}(f)$ *has the same value for any* $\psi^C = \lim_\nu \psi^C_{n_\nu}$ *in* (ii), *and accordingly,* $Q^A_{n_i}(f)$ *converges to* $I_{\psi^A}(f)$, *while* $Q^C_{n_i}(f)$ *converges to* $I_{\psi^C}(f)$. *The minimum of* $I_\psi(f)$, *is given, correspondingly, by* $I_{\psi^A}(f)$ *or* $I_{\psi^C}(f)$.

THEOREM 4.9. *In the Stieltjes case the sequence* $\left\{\dfrac{Z_{2n}(z)}{N_{2n}(z)}\right\}$ *of even approximants and the sequence* $\left\{\dfrac{Z_{2n-1}(z)}{N_{2n-1}(z)}\right\}$ *of odd approximants of the corresponding continued fraction* $\mathcal{C}(z)$ *both converge at any point* $z$, *real or complex, not on* $[0, \infty)$. *In other words, both sequences* $\{\psi^A_n\}$, $\{\psi^C_n\}$ *converge.*

The proof of this fundamental result of Stieltjes [5] is a direct application of Theorem 4.8 to the function $f(t) = \dfrac{1}{t - z}$. First, let $z$ be real and negative. Then $f(t) \in (2, n)$ in $[0, \infty)$. Hence $\displaystyle\int_0^\infty \frac{d\psi}{z - t}$ has the same value for any $\psi^A(t)$, and the convergence of $\{\psi^A_n(t)\}$, which is equivalent to that of $\mathcal{A}(z)$, now follows directly. The statement concerning $\left\{\dfrac{Z_{2n-1}(z)}{N_{2n-1}(z)}\right\}$ follows, by applying similar considerations to $f(t) = \dfrac{1}{z - t}$ in $[0, \infty)$, $z < 0$. Theorem 4.9 now follows, by the Vitali Theorem.

In Theorem 4.8 take $f(t) = \dfrac{1}{z - t}$ and use (4.18). Since, by (4.46) and (4.47),

$$R^A_{n,\psi}\left(\frac{1}{z - t}\right) < 0, \qquad R^C_{n,\psi}\left(\frac{1}{z - t}\right) > 0, \qquad z < 0,$$

we get

THEOREM 4.10. *For any solution* $\psi(t)$ *and for any* $z < 0$,

$$\frac{Z_{2n-1}(z)}{N_{2n-1}(z)} < \int_0^\infty \frac{d\psi^C}{z - t} \leqq \int_0^\infty \frac{d\psi}{z - t} \leqq \int_0^\infty \frac{d\psi^A}{z - t} < \frac{Z_{2n}(z)}{N_{2n}(z)}.$$

This shows immediately that the Stieltjes moment problem is determined if and only if $\mathcal{C}(z)$ converges.

In the Stieltjes case we also have the following result.

If, for a certain $\{n_\nu\}$, $f^{(2n_\nu+2)}(t) \geqq 0$ and $f^{(2n_\nu+1)}(t) \leqq 0$ in $[0, \infty)$, then $Q_{n_\nu}(f) > I_{\psi^C}(f) \geqq I_\psi(f) > Q_n^A(f)$. In particular, this holds, $\{n_\nu\} \equiv \{n\}$, if $f(t)$ is completely monotonic in $[0, \infty)$.

**10.** We return to approximate quadratures. Theorem 4.6, combined with the example (4.49), shows that (i) $Q_n^A(f)$ *based on the roots of Laguerre polynomials* $L_n^{(\alpha)}(x)$ *converges, if* $f(t) = O\left(\dfrac{e^t}{t^\alpha \log^{1+\rho_1} t \log_2^{1+\rho_2} t \cdots \log_m^{1+\rho_m} t}\right)$, *as* $t \to \infty$; $m \geqq 1$; $\rho_1, \rho_2, \cdots, \rho_m > 0$. (ii) $Q_n^A(f)$ *based on the roots of Hermite polynomials* $H_n(x)$ *converges, if* $f(t) = O\left(\dfrac{e^{t^2}}{|t| \log^{1+\rho_1} |t| \cdots \log_m^{1+\rho_m} |t|}\right)$.

This includes the results of Uspensky [2], as well as those of Jouravsky [1], obtained in a different manner. In order to obtain more general results we need

LEMMA 4.7. *Let* $f(t) = \sum_{n=0}^{\infty} c_n t^n$ *be an integral transcendental function with non-negative coefficients, and let* $\psi(t)$ *be an arbitrary solution of the Stieltjes moment problem. A necessary and sufficient condition for the existence of* $\int_0^\infty f(t)\, d\psi$ *is the convergence of* $\sum_{n=0}^{\infty} \mu_n c_n$. *Moreover,* $\int_0^\infty f(t)\, d\psi = \sum_{n=0}^{\infty} \mu_n c_n$, *hence, does not depend on the choice of* $\psi$. *Analogous results hold in* $(-\infty, \infty)$ *for* $f(t) = \sum_{n=0}^{\infty} c_n t^{2n}$, $c_n \geqq 0$, *and the series* $\sum_{n=0}^{\infty} \mu_{2n} c_n$.

Consider sections $f_n(t) \equiv c_0 + c_1 t + \cdots + c_n t^n$, $n = 0, 1, \cdots$. Evidently, $f_n(t) \to f(t)$, increasingly with respect to $n$, for any $t$ in $[0, \infty)$. Hence, by Lebesgue's Theorem,

$$\sum_{n=0}^{\infty} \mu_n c_n = \lim_n \int_0^\infty f_n(t)\, d\psi = \int_0^\infty \lim_n f_n(t)\, d\psi = \int_0^\infty f(t)\, d\psi .$$

The proof for $(-\infty, \infty)$ is quite similar.

COROLLARY 4.6. *In the Stieltjes case,* $Q_n^A(f)$ *and* $Q_n^C(f)$ *both converge for any integral transcendental function* $f(t) = \sum_{n=0}^{\infty} c_n t^n$ *with non-negative coefficients, for which* $\sum_{i=0}^{\infty} \mu_i c_i$ *converges, and*

$$\lim_n Q_n^A(f) = \lim_n Q_n^C(f) = \sum_{i=0}^{\infty} \mu_i c_i = \int_0^\infty f(t)\, d\psi.$$

*Furthermore, if, as $t \to \infty$, $f_0(t)$ is dominated by a function $f(t)$ of the above type, then*

$$Q_n^A(f_0) \to \int_0^\infty f_0(t)\, d\psi^A, \qquad Q_n^C(f_0) \to \int_0^\infty f_0(t)\, d\psi^C.$$

For the interval $(-\infty, \infty)$ Jouravsky[1] has obtained a convergence criterion under the following conditions

$$\mu_{2n} = O[(2n+p)!\, l^n], \qquad l, p > 0 \text{ fixed}, \qquad \psi(t) \text{ absolutely continuous}$$

(here the moment problem is determined, by Carleman's criterion):

THEOREM 4.11. *Let $f(t)$ be an integral transcendental function $\sum_{n=0}^{\infty} c_n t^n$, $c_n \geqq 0$, and, as $|t| \geqq \infty$, let*

$$f_0(t) = O\left[\frac{f(t^2)}{|t|^s + \lambda^2}\right],$$

*$s = 2k + 2\sigma$, $k$ a positive integer, $0 \leqq \sigma < 1$. Assume further that $\sum_{n=0}^{\infty} \mu_{2n-2}^{\sigma} \mu_{2n}^{1-\sigma} c_{n+k}$ converges. Then $f_0(t) \in F^A$.*

We pass now to the Hamburger case. Here we cannot go as far as previously. In fact, although we know that $\{\psi_n^C(t)\}$ converges, we do not have an adequate expression for $R_{n,\psi}^C(f)$; on the other hand, an adequate expression is available for $R_{n,\psi}^A(f)$, but $\{\psi_n^A(t)\}$ does not converge in the general case.

The following results of a more particular character are derived from the preceding discussion, by reasoning quite similar to that developed in the Stieltjes case.

(i) If $f(t) = \sum_{n=0}^{\infty} c_n t^{2n}$ is an integral transcendental function with non-negative coefficients, such that $\sum_{i=0}^{\infty} \mu_{2i} c_i$ converges, then $Q_n^A(f)$ converges to $\sum_{i=0}^{\infty} \mu_{2i} c_i = \int_{-\infty}^{\infty} f(t)\, d\psi$, where $\psi(t)$ is an arbitrary solution of the moment problem.

Indeed, $I_\psi(f)$ exists and $f(t) \in (2, n)$, hence, $Q_n^A(f)$ is bounded. If now $\psi^A(t) = \lim_{\nu} \psi_{n_\nu}^A(t)$, then, by Lemmas 4.6 and 4.7, $Q_{n_\nu}^A(f)$ converges to the value $\sum_{i=0}^{\infty} \mu_{2i} c_i = \int_{-\infty}^{\infty} f(t) d\psi$.

(ii) If $f_0(t)$, as $|t| \to \infty$, is dominated by $f(t)$ of (i), then $Q_{n_\nu}^A(f_0) \to I_{\psi^A}(f_0)$, where $\{\psi_{n_\nu}^A(t)\}$ is a convergent subsequence, with $\psi^A(t)$ as its limit function.

**11.** Approximate quadrature formulas give rise to an interesting application to the following problem: *how wide can be an interval known to be free of roots $\{x_{jn}\}$, for a given $n$?*

Consider, first, the case of a finite orthogonality interval $[a, b]$. Assume that $(c, d) \subset [a, b]$ contains no roots of $q_n(x)$. Define

$$(4.50)\qquad f(x) = \begin{cases} (x - c)^k(d - x)^k \text{ in } [c, d], \\ 0 \text{ elsewhere,} \end{cases}$$

where $k$ is a positive integer to be specified later. We readily verify the following relations:

$$(4.51)\qquad \begin{cases} |f(x)| \leqq f\left(\dfrac{c+d}{2}\right) = \left(\dfrac{d-c}{2}\right)^{2k} = \gamma^{2k}, \qquad \gamma = \dfrac{d-c}{2}, \qquad a \leqq x \leqq b. \\ |f^{(k)}(x)| \leqq 2^k \gamma^k \cdot k!. \end{cases}$$

The last inequality is derived by means of elementary properties of Legendre polynomials. Hence [Jackson, 1] there exists a polynomial $P_n(x)$, of degree $n > k - 1$, such that in $a \leqq x \leqq b$

$$(4.52)\qquad |f(x) - P_n(x)| \leqq \left[\frac{2\mathfrak{L} k\gamma(b-a)}{n}\right]^k,$$

where $\mathfrak{L}$ is an absolute constant. Moreover, a simple computation shows that $|f'(x)|$ is maximum at

$$\xi_1 = \frac{c+d}{2} - \frac{\gamma}{\sqrt{2k-1}}, \qquad \xi_2 = \frac{c+d}{2} + \frac{\gamma}{\sqrt{2k+1}}.$$

Consider now the subinterval

$$I: \left[\frac{c+d}{2} - \frac{\gamma}{4k}, \qquad \frac{c+d}{2} + \frac{\gamma}{4k}\right]$$

or one half of it if $(c - a)(d - b) = 0$. The points $\xi_1$, $\xi_2$ certainly fall outside $I$ for $k$ sufficiently large. Hence,

$$|f'(x)| \leqq f'(x_0) \leqq \frac{\gamma^{2k-1}}{2}, \qquad x_0 = \frac{c+d}{2} - \frac{\gamma}{4k}, \qquad x \in I,$$

$$(4.53)\qquad 0 < f\left(\frac{c+d}{2}\right) - f(x_0) = \gamma^{2k} - f(x_0) \leqq \int_{x_0}^{\frac{c+d}{2}} f'(x_0)\, dx \leqq \frac{\gamma^{2k}}{8k}$$

$$f(x) \geqq f(x_0) > \gamma^{2k}\left(1 - \frac{1}{8k}\right) > \frac{3}{4}\gamma^{2k}, \qquad x \in I.$$

Apply now the approximate quadrature formula (4.4) to the function (4.50). We have

$$(4.54)\qquad \begin{aligned} R_{n,\psi}(f) &= \int_c^d f(x)\, d\psi = R_{n,\psi}(f - P_n) \\ &= \int_a^b [f(x) - P_n(x)]\, d\psi - \int_a^b [f(x) - P_n(x)]\, d\psi_n. \end{aligned}$$

By (4.51), (4.52) and (4.53),

$$\frac{3}{4}\gamma^{2k}\int_I d\psi \leqq \int_I f(x)\,d\psi \leqq 2\mu_0\left[\frac{2\mathfrak{L}k\gamma(b-a)}{n}\right]^k, \qquad n > k-1. \tag{4.55}$$

Assume now

$$\int_r^{r+l} d\psi \geqq ml, \tag{4.56}$$

where $[r, r+l]$ is any subinterval of $[c, d]$ and $m > 0$ does not depend on $r$, $l$.

This is certainly satisfied, if

$$d\psi(x) = h(x)\,dx, \qquad h(x) \geqq m > 0 \quad \text{in} \quad (c, d). \tag{4.57}$$

Taking in (4.55) $k \sim \log n$, we get the desired result

$$d - c < \frac{\tau \log n}{n}.$$

Here, and in the subsequent discussion, we generally denote by $\tau$ a properly fixed positive quantity, independent of $c$, $d$, $n$.

The above analysis is applicable to any formula of approximate quadrature, with a positive $d\psi(x)$, which is exact for polynomials, provided the sum of the absolute values of its coefficients remains bounded with respect to $n$. It follows that if $x_{i-1,n}$, $x_{i,n}$ both lie in an interval $[k, k_1]$ where (4.56) is satisfied, then

$$x_{i,n} - x_{i-1,n} < \frac{\tau \log n}{n}. \tag{4.58}$$

By means of Tchebycheff's inequalities (4.36), this yields

$$\rho_{in} < \frac{\tau \log n}{n}, \qquad \text{if} \quad x_{i-1,n}, x_{i,n}, x_{i+1,n} \quad \text{all lie in} \quad [k, k_1]. \tag{4.59}$$

If $x_{i-2,n}$, $x_{i-1,n}$, $x_{i,n}$, $x_{i+1,n}$ are all in $[k, k_1]$ and $x_{i-1,n} \leqq \alpha \leqq x_{i,n}$, then for any two solutions $\psi'(x)$, $\psi''(x)$ of the reduced moment problem (2.9)

$$\begin{cases} \left| \int_a^\alpha d\psi'(x) - \sum^{[a,\alpha]} \rho_{jn} \right| < \dfrac{\tau \log n}{n} \\ \left| \int_a^\alpha d\psi'(x) - \int_a^\alpha d\psi''(x) \right| < \dfrac{\tau \log n}{n}. \end{cases} \tag{4.60}$$

The above considerations have many points in common with those of [Krawtchouk [1, 3] (cf. also [Netzorg, 1]). They apply to a general class of quasi-orthogonal polynomials, while finer results may be obtained in special cases. Thus, if $d\psi(x) = h(x)\,dx$, where $0 < m \leqq h(x) \leqq M < \infty$ in $[a, b]$, then, according to Erdös and Turán, [1], we have, for $[a, b] = [-1, 1]$,

$$\theta_{jn} - \theta_{j-1,n} < \frac{\tau}{n}; \qquad \xi_{jn} = \cos\theta_{jn}, \qquad j = 1, 2, \cdots, n.$$

Estimates of $\xi_{jn} - \xi_{j-1,n}$ based on the asymptotic expression for the corresponding orthogonal polynomials are given by Szegö [2]. The case of the interval

$(-\infty, \infty)$ may be treated in a similar manner. Again introduce the function $f(x)$ as in (4.50) and an interval $[-A_n, A_n]$ increasing indefinitely with $n$ and such that $[-A_n, A_n] \supset [x_{0n}, x_{nn}]$. There exists a polynomial $P_n(x)$, of degree $n > k - 1$, such that

$$| f(x) - P_n(x) | \leqq \left[\frac{4\mathfrak{L}k\gamma A_n}{n}\right]^k, \qquad -A_n \leqq x \leqq A_n,$$

whence

$$(4.61) \qquad | P_n(x) | \leqq l_k \gamma^k \text{ in } [d, A_n] \text{ and } [-A_n, c], \qquad l_k \equiv \left(\frac{4\mathfrak{L}kA_n}{n}\right)^k.$$

We now apply to the polynomial $P_n(x)$ the following inequality of Tchebycheff: $| P_n(x) | \leqq M$ on $[\alpha, \beta]$ implies, for $\xi$ outside $[\alpha, \beta]$,

$$| P_n(x) | \leqq \frac{M}{2} \left| \frac{[2\xi - \alpha - \beta + 2\sqrt{(\xi - \alpha)(\xi - \beta)}]^n + [2\xi - \alpha - \beta - 2\sqrt{(\xi - \alpha)(\xi - \beta)}]^n}{(\beta - \alpha)^n} \right|.$$

Using (4.61), we readily get

$$| P_n(x) | < l_k \gamma^k \left(\frac{4}{A_n - d}\right)^n | x |^n, \qquad | x | \geqq A_n, \qquad 0 \leqq c < d < A_n.$$

For $R_{n,\psi}(f)$ we now write a relation analogous to (4.54), which yields

$$0 < R_{n,\psi}(f) = \int_c^d f(x)\, d\psi \leqq l_k \gamma^k (2\mu_0 + d_n),$$

$$d_n = \left(\frac{4}{A_n - d}\right)^n \left[\int_{-\infty}^{-A_n} (-x)^n\, d\psi + \int_{A_n}^{\infty} x^n\, d\psi\right].$$

Assume that (4.56) is satisfied in any subinterval of $[c, d]$, where now $c, d, m$ eventually depend on $n$. By a reasoning quite similar to that employed for a finite interval, we get

$$d - c < \tau k \left(\frac{A_n}{n}\right)^{\frac{k}{k+1}} m^{-\frac{1}{k+1}} (2\mu_0 + d_n)^{\frac{1}{k+1}},$$

whence, taking $k \sim \log n$ and assuming that $m$ and $d_n$ are bounded with respect to $n$,

$$(4.62) \qquad d - c < \tau \frac{A_n}{n} \log n.$$

The inequalities (4.58), (4.59) and (4.60) now follow, with $\dfrac{\log n}{n}$ replaced by $\dfrac{A_n}{n} \log n$.

**12.** We refer to [Krawtchouk 8, 9] for a discussion of an upper bound for $\left| \int_a^x d\psi'(t) - \int_a^x d\psi''(t) \right|$, where $\psi'$ and $\psi''$ are two distribution functions whose first $N$ moments are "approximately" equal.

# BIBLIOGRAPHY.


**Achyeser (Achieser), N. I.**

1. *Infinite Jacobi Matrices and the Problem of Moments* (in Russian). Uspekhi Matematicheskikh Nauk, 9 (1941), 126–156.

**Achyeser (Achieser), N. I. und Krein, M.**

1. *Ueber Fouriersche Reihen beschränkter summierbarer Funktionen und ein neues Extremumproblem.* I, Communications de la Société Mathématique de Kharkoff, (4) 9 (1934), 3–28; II, *ibid.*, (4) 10 (1934), 3–32.

2. *Ueber eine Transformation des reellen Toeplitzschen Formen und das Momentenproblem in einem endlichen Intervalle.* Communications de la Société Mathématique de Kharkoff, 11 (4) (1935), 21–26.

3. *Das Momentenproblem bei der zusätzlicher Bedingung von A. Markoff.* Communications de la Société Mathématique de Kharkoff, 12 (4) (1935), 13–36.

4. *Sur deux questions de minima qui se rattachent au problème des moments.* Comptes Rendus de l'Académie des Sciences de l'URSS, (2) 4 (1) (1936), 343–346.

5. *The problem of moments on two intervals under the condition of A. Markoff.* Communications de la Société Mathématique de Kharkoff, 14 (4) (1937), 47–60.

6. *On certain problems in the theory of moments* (in Russian). Kharkoff, 1938, 254 pp.

7. *Some remarks about three papers of M. S. Verblunsky.* Communications de l'Institut des Sciences Mathématiques et Mécaniques de l'Université de Kharkoff et de la Société Mathématique de Kharkoff, (4), 16 (1940), 129–134.

**Adamoff, A.**

1. *Démonstration d'une proposition de Stieltjes.* Bulletin de la Société Physico-Mathématique de Kazan, (2) 11 (1902), 1–12.

**Bailey, R. P.**

1. *Convergence of sequences of positive linear functional operations.* Duke Mathematical Journal, 2 (1937), 285–303.

**Banach, S.**

1. *Théorie des opérations linéaires.* Warszawa, 1932, 254 pp.

**Bateman, H.**

1. *On the inversion of a definite integral.* Proceedings of the London Mathematical Society, 4, (2), (1907), 461–498.

**Bergman, S. and Martin, W. T.**

1. *A modified moment problem in two variables.* Duke Mathematical Journal, 6 (1940), 389–407.

**Bernstein, F.**

1. *Die Uebereinstimmung derjenigen beiden Summationsverfahren einer divergenten Reihe welche von T. J. Stieltjes und E. Borel herrühren.* Jahresbericht der deutschen Mathematiker-Vereinigung, 28 (1919), 50–63.

2. *Berechtigung zu der Arbeit: Die Uebereinstimmung derjenigen beiden Summationsverfahren einer divergenten Reihe, welche von T. J. Stieltjes und E. Borel herrühren.* Jahresbericht der deutschen Mathematiker-Vereinigung, 29 (1920), 94.

3. *Bermerkung zu der vorstehenden Abhandlung: Ueber die Konvergenz eines mit einer Potenzreihe associierten Kettenbruches, von H. Hamburger in Berlin.* Mathematische Annalen, 81 (1920), 46–47.

**Bernstein, S.**

1. *On the best approximation of continuous functions by polynomials of a given degree* (in Russian). Kharkoff, 1912, 153 pp.

2. *Sur les fonctions absolument monotones.* Acta Mathematica, 52 (1929), 1–66.

**Blumenthal, O.**

1. *Ueber die Entwicklung einen willkürlichen Funktion nach den Nennern der Kettenbrüches für* $\int_{-\infty}^{0} \frac{\varphi(\xi)d\xi}{z-\xi}$. Thesis, Göttingen, 1898, 57 pp.

**Boas, R. P.**

1. *Necessary and sufficient conditions in the moment problem for a finite interval.* Duke Mathematical Journal, 1 (1935), 449–476.

2. *On a generalization of the Stieltjes moment problem.* Transactions of the American Mathematical Society, 46 (1939), 142–150.

3. *The Stieltjes moment problem for functions of bounded variation.* Bulletin of the American Mathematical Society, 45 (1939), 399–404.

4. *A trigonometric moment problem.* Journal of the London Mathematical Society, 14 (1939), 242–244.

5. *A general moment problem.* American Journal of Mathematics, 63 (1941), 361–370.

**Bochner, S.**

1. *Monotone Funktionen, Stieltjessche Integrale und harmonische Analyse.* Mathematische Annalen, 108 (1933), 378–410.

2. *Ueber Funktionen mit n Zeichenwechseln, deren Momente bis zur Ordung n-1 verschwinden.* Jahresbericht der Deutschen Mathematiker-Vereinigung, 50 (1940), 129–139.

**Bödewadt, U. T.**

1. *Zum Momentproblem für das Interval* (0, 1). Mathematische Zeitschrift, 40 (1935), 426–462.

**Borel, E.**

1. *Sur quelques points de la théorie des fonctions.* Annales de l'Ecole Normale Supérieure, (3) 12 (1895), 9–55.

**Cantelli, F. P.**

1. *Estensioni del teorema di reciprocita di Fourier, sue relazioni col problema dei momenti e coi piu recenti progressi del calcolo delle probabilita.* Acti della Societa Italiana per il progresso delle scienze, 18 (2) (1929), 25–30.

**Carleman, T.**

1. *Sur le problème des moments.* Comptes Rendus, 174 (1922), 1680–1682.

2. *Sur les équations intégrales singulières a noyau réel et symétrique.* Uppsala Universitets Årsskrift, 1923, 228 pp.

3. *Les fonctions quasi-analytiques.* Gauthier-Villars, Paris, 1926, 115 pp.

**Castelnuovo, G.**

1. *Sul problema dei momenti.* Giornale dell'Istituto Italiano degli Attuari, 1 (1930), 3–35.

2. *Sul problema dei momenti.* Giornale dell'Istituto Italiano degli Attuari, 1 (1930), 137–139.

3. *Sur quelques problèmes se rattachant au Calcul des Probabilités.* Annales de l'Institut Henri Poincaré, 3 (1933), 465–490.

**Chlodowsky, I.**

1. *Les fonctions presque absolument monotones.* Comptes Rendus (Doklady) de l'Académie des Sciences de l'URSS, 25 (1939), 727–730.

**Christoffel, E. B.**

1. *Sur une classe particulière de fonctions entières et de fractions continues.* Annali di Mathematica, 8 (2) (1877), 1–10.

**Cramér, H.**

1. *Random variables and probability distributions.* Cambridge Tracts in Mathematics and Mathematical Physics, No. 36, Cambridge, 1937, 121 pp.

**de Montessus de Ballore, R.**

1. *Sur les fractions continues algébriques.* Comptes Rendus, 134 (1902), 1489–1491.

2. *Sur les fractions continues algébriques de Laguerre.* Comptes Rendus, 140 (1905), 1438–1440.

3. *Les fractions continues algébriques.* Acta Mathematica, 32 (1909), 257–281.

**Dominguez, A. G.**

1. *Sur les intégrales de Laplace.* Comptes Rendus, 205 (1937), 1035–1038.

**Dusl, K.**

1. *Quelques remarques sur les polynomes généralisés de Laguerre.* Verhandlungen des internationalen Mathematiker-Kongresses Zurich, 2 (1932), 136–139.

**Edrei, A.**

1. *Sur les déterminants recurrents et sur les singularités d'une fonction donnée par son développement de Taylor.* Compositio Mathematica, 7 (1940), 20–88.

**Erdelyi, A.**

1. *Ueber eine erzeugende Funktion von Produkten Hermitescher Polynome.* Mathematische Zeitschrift, 44 (1938), 201–211.

**Erdös, P. and Turan, P.**

1. *On Interpolation.* II. *On the distribution of the fundamental points of Lagrange and Hermite interpolation.* Annals of Mathematics, 39 (1938), 703–724.

**Evans, G. C.**

1. *The logarithmic potential.* American Mathematical Society Colloquium Publications, 6, New York, 1927, 150 pp.

**Faber, A.**

1. *Ueber nach Polynomen fortschreitende Reihen.* Sitzungsberichte der mathematisch-physikalischen Classe der bayerischen Akademie der Wissenschaften zu München, (1922), 157–178.

**Favard, J.**

1. *Sur les zéros réels des polynomes.* Bulletin de la Société Mathématique de France, 59 (1931), 229–255.

2. *Sur les polynomes de Tchebichef.* Comptes Rendus, 200 (1935), 2052–2055.

**Fejér, L.**

1. *Mechanische Quadraturen mit positiven Cotes'schen Zahlen.* Mathematische Zeitschrift, 37 (1933), 287–309.

**Feldheim, E.**

1. *Sur l'orthogonalité des fonctions fondamentales, et sur la forte convergence en moyenne des polynomes d'interpolation de Lagrange dans le cas des abscisses de Tchebycheff.* Bulletin de la Société Mathématique de France, 65 (1937), 1–40.

2. *Sur le mode de convergence dans l'interpolation de Lagrange.* Comptes Rendus de l'Académie des Sciences de l'URSS, (2) 5 (1) (1937), 327–331.

3. *Théorie de la convergence des procédés d'interpolation et de quadrature mécanique.* Mémorial des Sciences Mathématiques, Fasc. 95 (1939), 90 pp.

**Fischer, E.**

1. *Ueber das Caratheodorysche Problem, Potenzreihen mit positivem reellen Teil betreffend.* Rendiconti del Circolo Matematico di Palermo, 32 (1911), 240–256.

**Fréchet, Maurice.**

1. *Généralités sur les probabilités. Variables aléatoires.* Traité du Calcul des Probabilités et de ses Applications par E. Borel. T. I, Fasc. III, Paris, Gauthier-Villars, 1937, 308 pp.

**Fréchet, M. and Shohat, J.**

1. *A proof of the generalized second limit theorem in the theory of probability.* Transactions of the American Mathematical Society, 33 (1931), 533–543.

**Friedrichs, K. and Horvay, G.**

1. *The finite Stieltjes Momentum Problem.* Proceedings of the National Academy of Sciences, 25 (1939), 528–534.

**Garabedian, H. L. and Wall, H. S.**

1. *Hausdorff methods of summation and continued fractions.* Transactions of the American Mathematical Society, 48 (1940), 185–207.

**Garabedian, H. L., Hille, E., and Wall, H. S.**

1. *Formulation of the Hausdorff inclusion problem.* Duke Mathematical Journal, 8 (1941), 193–213.

**Geronimus, J.**

1. *On some problems involving the persymmetric determinants.* Proceedings of the Royal Society of Edinburgh, 51 (1931), 14–18.

2. *On some persymmetric determinants formed by the polynomials of M. Appell.* The Journal of the London Mathematical Society, (6) (1931), 55–59.

3. *On polynomials orthogonal with regard to a given sequence of numbers.* Comm. Inst. Sci. Math. Mec. Univ. Kharkoff (Zapiski Inst. Mat. Mech.) (4) 17, 3–18 (1940). (Russian, English summary.)

4. *Sur quelques équations aux différences finies et les systemes correspondants des polynomes orthogonaux.* Comptes Rendus (Doklady) de l'Académie des Sciences de l'URSS, 29 (1940), 536–538.

5. *On some properties of the generalized orthogonal polynomials* (in Russian). Mathematicheski Zbornik, 9 (1941), 121–135.

6. *On the character of the solution of the moment-problem in the case of the periodic in the limit associated fraction* (in Russian). Bulletin de l'Académie des Sciences de l'URSS, 5 (1941), 203–210.

**Giraud, G.**

1. *Sur deux formules applicables au calcul numérique des intégrales.* Comptes Rendus, 178 (1924), 2227–2229.

**Gourier, G.**

1. *Sur une méthode capable de fournir une valeur approchée de l'intégrale* $\int_{-\infty}^{\infty} F(x)dz$. Comptes Rendus, 97 (1883), 79–82.

**Grommer, J.**

1. *Ganze transzendente Funktionen mit lauter reellen Nullstellen.* Journal für die reine und angewandte Mathematik, 144 (1914), 114–166.

**Hallenbach, F.**

1. *Zur Theorie der Limitierungsverfahren von Doppelfolgen.* Thesis (Bonn a. Rh.), 1933, 100 pp.

**Hamburger, H.**

1. *Beiträge zur Konvergenztheorie der Stieltjes'schen Kettenbrüche.* Mathematische Zeitschrift, 4 (1919), 186–222.

2. *Ueber die Konvergenz eines mit einer Potenzreihe assoziierten Kettenbruchs.* Mathematische Annalen, 81 (1920), 31–45.

3. *Ueber eine Erweiterung des Stieltjes'schen Momentenproblems.* Mathematische Annalen, 81 (1920), 235–319; 82 (1921), 120–164, 168–187.

**Hardy, G. H.**

1. *On Stieltjes' "problème des moments".* Messenger of Mathematics, a) 46 (1917), 175–182; b) 47 (1917), 81–88.

**Hausdorff, F.**

1. *Summationsmethoden und Momentfolgen.* Mathematische Zeitschrift, 9(1921); I, 74–109; II, 280–299.

2. *Momentprobleme für ein endliches Intervall.* Mathematische Zeitschrift, 16 (1923), 220–248.

**Haviland, E. K.**

1. *On Distribution Functions and their Laplace-Fourier Transform.* Proceedings of the National Academy of Sciences, 20 (1934), 50–57.

2. *On the theory of absolutely additive distribution functions.* American Journal of Mathematics, 56 (1934), 625–657.

3. *On the inversion formula for Fourier-Stieltjes transforms in more than one dimension.* American Journal of Mathematics, 57 (1935), 94–100; II, 382–388.

4. *On the momentum problem for distributions in more than one dimension.* American Journal of Mathematics, 57 (1935), 562–568.

5. *On the momentum problem for distribution functions in more than one dimension.* II. American Journal of Mathematics, 58 (1936), 164–168.

**Haviland, E. K. and Wintner, A.**

1. *On the Fourier-Stieltjes transform.* American Journal of Mathematics, 56 (1934), 1–7.

**Hedge, L. B.**

1. *Moment problem for a bounded region.* Bulletin of the American Mathematical Society, 47 (1941), 282–285.

**Hellinger, E.**

1. *Zur Stieltjes'schen Kettenbruchtheorie.* Mathematische Annalen, 86 (1922), 18–29.

**Hellinger, E. and Toeplitz, E.**

1. *Zur Einordnung der Kettenbruchtheorie in die Theorie der quadratischen Formen von unendlichvielen Veränderlichen.* Journal für die reine und angewandte Mathematik, 144 (1914), 212–238.

**Herglotz, A.**

1. *Ueber Potenzreihen mit positiven reellen Teil im Einheitskreise.* Berichte über die Verhandlungen der (königlich) Sächsischen Gesellschaft (Akademie) der Wissenschaften zu Leipzig, mathematisch-physische Klasse, 63 (1911).

**Hermite, C.**

1. *Correspondence d'Hermite et de Stieltjes*, 2 (329) (1892), 232–234; 2 (337) (1892), 249–250.

**Hildebrandt, T. H.**

1. *On the moment problem for a finite interval.* Bulletin of the American Mathematical Society, 38 (1932), 269–270.

**Hildebrandt, T. H. and Schoenberg, I. J.**

1. *On linear functional operators and the moment problem for a finite interval in one or several dimensions.* Annals of Mathematics, (2) 34 (1933), 317–328.

**Hille, E.**

1. *Bilinear formulas in the theory of the transformation of Laplace.* Compositio Mathematica, 6 (1938), 93–102.

**Hille, E. and Tamarkin, J.**

1. *Questions of relative inclusion in the domain of Hausdorff means.* Proceedings of the National Academy of Sciences, 19 (1933), 573–577.

2. *On moment functions.* Proceedings of the National Academy of Sciences, 19 (1933), 902–908.

**Horvay, G. and Friedrichs, K.** (See Friedrichs, K.)

**Jackson, D.**

1. *The theory of approximation.* American Mathematical Society Colloquium Publications, 11, New York, 1930, 178 pp.

**Jouravsky, A.**

1. *Sur la convergence des formules des quadratures mécaniques dans un intervalle infini.* Journal de la Société Physico-Mathématique de Leningrad, 2 (1928), 31–52.

**Kaczmarz, S. und Steinhaus, H.**

1. *Theorie des orthogonalreihen*, Warszawa-Lwow, 1935, 298 pp.

**Kantorovitch, L.**

1. *On the moment problem for a finite interval.* Comptes Rendus de l'Académie des Sciences de l'URSS, (2) 5 (1) (1937), 531–537.

**Kapteyn, W.**

1. *Ontwikkeling eenes functie neav de functies $\varphi_n$ van Abel.* K. Akademie van Wetenschappen te Amsterdam. Verslagen van de gewone Vergaderingen der wis—en Natuurkundige Afdeeling, 21 (2) (1913), 1204–1215.

2. *Expansion of a function in series of Abel's functions $\varphi_n(x)$.* K. Akademie van Wetenschappen te Amsterdam. Proceedings of the Section of Science, 15 (2) (1913), 1245–1256.

3. *Over de functies van Hermite.* K. Akademie van Wetenschappen te Amsterdam. Verslagen van de gewone Zergaderingen der wis—en Natuurkundige Afdeeling, 22 (2) (1914), 1057–1069, 1285–1294; 23 (1) (1914), 49–59.

4. *On Hermite's functions.* K. Akademie van Wetenschappen te Amsterdam. Proceedings of the Section of Science, 16 (2) (1914), 1191–1211; 17 (1) (1914), 139–150.

**Kendall, M. A.**

1. *Conditions for uniqueness in the problem of moments.* Annals of Mathematical Statistics, 11 (1940), 402–409. Correction: ibid., 12 (1941), 404–405.

**Korkine, A. et Zolotareff, A.**

1. *Sur un certain minimum.* Nouvelles Annales de Mathématiques, (2) 12 (1873), 337-355.

**Krawtchouk, M.**

1. *Sur la distribution des racines des polynomes orthogonaux.* Comptes Rendus, 196 (1933), 739–741.

2. *Quelques résultats nouveaux dans la méthode des moments.* Journal de l'Institut Mathématique de l'Académie des Sciences d'Ukraine, 1 (1933), 1–20.

3. *Sur le problème des moments.* Journal de l'Institut Mathématique de l'Académie des Sciences d'Ukraine, (1) (1933), 21–39.

4. *On a certain algebraic problem in the problem of moments.* Journal de l'Institut Mathématique de l'Académie des Sciences d'Ukraine, (1934), 87–92.

5. *Quelques conséquences de l'inégalité de M. Lebesgue.* Journal de l'Institut Mathématique de l'Académie des Sciences d'Ukraine, (1) (1935), 31–42.

6. *Sur le problème des moments.* Journal de l'Institut Mathématique de l'Académie des Sciences d'Ukraine, (2) (1935), 13–34.

7. *Sur une inégalité dans le problème des moments.* Journal de l'Institut Mathématique de l'Académie des Sciences d'Ukraine, (2) (1935), 35–43.

8. *Sur le problème des moments généralisés.* Journal de l'Institut Mathématique de l'Académie des Sciences d'Ukraine, (4) (1936), 3–8.

9. *Sur quelques approximations dans le problème des moments pour les fonctions de deux variables.* Comptes Rendus de l'Académie des Sciences de l'URSS, (7) 17 (1937), 283-285.

**Krein, M.**

1. *Ueber eine neue Klasse von Hermiteschen Formen und über eine Verallgemeinerung des trigonometrischen Momentenproblems.* Bulletin de l'Académie des Sciences de l'URSS, (7) 6 (1933), 1259-1275.

2. *On the spectrum of a Jacobian form in connection with the theory of vibrations of shafts* (in Russian). Mathematicheski Zbornik, 40 (1933), 455–466.

3. *Sur le problème de prolongement des fonctions hermitiennes et positives.* Comptes Rendus (Doklady) de l'Académie des Sciences de l'URSS, 26 (1940), 17–22.

**Krein, M. und Achyeser, N.** (see Achyeser)

**Kryloff, N.**

1. *Sur quelques formules d'approximation, fondées sur la généralisations des quadratures, dites mécaniques.* Comptes Rendus, 168 (1919), 721–723.

2. *Sur quelques recherches dans le domaine de la théorie de l'interpolation et des quadratures, dites mécaniques.* Proceedings of the International Mathematical Congress, Toronto, 1 (1924), 654–656.

**Laguerre, E.**

1. *Sur l'intégrale* $\int_x^\infty e^{-x}x^{-1}\,dx$. Bulletin de la Société Mathématique de France, 7 (1878), 72-81.

**Le Roy, E.**

1. *Sur les séries divergentes et les fonctions définies par un développement de Taylor.* Comptes Rendus, 127 (1898), 654–657.

2. *Sur les séries divergentes et les fonctions définies par un développement de Taylor.* Annales de la Faculté des Sciences de Toulouse, (2) 2 (1900), 317–430.

**Lifshetz, M.**

1. *On some questions concerning the determinate case of Hamburger's moment problem* (in Russian). Mathematicheski Zbornik, 6 (48) (1939), 293-306.

**Maillet, E.**

1. *Sur les fractions continues algébriques.* Journal de l'Ecole Polytechnique, (2) 12 (1907), 41-63.

**Markoff, A.**

1. *On certain applications of algebraic continued fractions.* Thesis (in Russian), St. Petersburg, 1884, 131 pp.

2. *Démonstration de certaines inégalités de Tchebychef.* Mathematische Annalen, 24 (1884), 172–180.

3. *Sur la méthode de Gauss pour le calcul approché des intégrales.* Mathematische Annalen, 25 (1885), 427–432.

4. *Extrait d'une lettre addressée a M. Hermite.* Annales Scientifiques de l'Ecole Normale Supérieure, 3 (3) (1886), 81–88.

5. *Sur une question de maximum et de minimum proposée par M. Tchebycheff.* Acta Mathematica, 9 (1886), 57–70.

6. *Proof of convergence of many continued fractions* (in Russian). Zapiski Imperatorskoi Akademii Nauk (St. Petersburg) (1893), 8 pp. (Supplement)

7. *On functions generated by developing power series in continued fractions* (in Russian). Zapiski Imperatorskoi Akademii Nauk (St. Petersburg) (Supplement) (1894), 1–30. English translation, by J. Shohat, Duke Mathematical Journal, 7 (1940), 85–96.

8. *Note sur les fractions continues.* Bulletin de la classe Physico-Mathématique de l'Académie Impériale des Sciences de Saint-Petersburg, 5 (2) (1895), 9–13.

9. *Deux démonstrations de la convergence de certaines fractions continues.* Acta Mathematica, 19 (1895), 93–104.

10. *Nouvelles applications des fractions continues.* Mémoires de l'Académie des Sciences de St.-Petersburg, classe physico-mathématique, (8) 3 (5) (1896), 50 pp.

11. *Nouvelles applications des fractions continues.* Mathematische Annalen, 47 (1896), 579–597.

12. *Differenzenrechnung.* Teubner, Leipzig, 1896.

13. *Sur les valeurs extrèmes des intégrales et l'interpolation.* Mémoires de l'Académie des Sciences de St.-Petersburg, classe physico-mathématique, (8) 6 (5) (1898), 69 pp.

14. *Recherches sur les valeurs limites des intégrales.* Mémoires de l'Académie des Sciences de St.-Petersburg, classe physico-mathematique, (8) 10 (9) (1900), 34 pp.

15. *Calculus of Probability* (in Russian). (Soviet Press), Moscow, 4th edition, 1924, 588 pp.

**Martin, W. T. and Bergman, S.** (see Bergman, S.)

**M'Arthur, N.**

1. *Note on the polynomials which satisfy the differential equation* $xy'' + (x - \gamma)y' - \alpha y = 0$. Proceedings of the Edinburgh Mathematical Society, 38 (1919), 27–33.

**von Mises, R. (de Misés, R.)**

1. *Fundamentalsätze der Wahrscheinlichkeitrechnung.* Mathematische Zeitschrift, 4 (1919), 1–97.

2. *The limits of a distribution function if two expected values are given.* Annals of Mathematical Statistics, 10 (1939), 99–104.

3. *An inequality for the moments of a discontinuous distribution.* Skand. Aktuarietidskrift (1939), 32–36.

4. *Bestimmung einer Verteilung durch ihre erste Momente.* Skand. Aktuarietidskrift (1937).

5. *Sur une inégalité pour les moments d' une distribution convexe.* Bulletin des Sciences Mathématiques, (2) 62 (1938), 68–71.

**Murphy, R.**

1. *On the inverse method of definite integrals, with physical applications.* Transactions of the Cambridge Philosophical Society, 4 (1833), 353–408; 5 (1835) 113–148, 315–393.

**Netzorg, D. L.**

1. *On mechanical quadrature formulas and roots of orthogonal polynomials.* Thesis, University of Illinois, 1937.

**Nevanlinna, R.**

1. *Asymptotische Entwickelungen beschränkter Funktionen und das Stieltjessche Momentenproblem.* Annales Academiae Scientiarum Fennicae, (A) 18 (5) (1922), 52 pp.

**Nielsen, N.**

1. *Recherches sur une classe des fonctions méromorphes.* Danske Videnskab. Selskabs Skrifter, natur. ok math., (7) 2 (2) (1904), 57–100.

**Obrechkoff, N.**

1. *Sur les zéros de quelques classes de polynomes.* Annuaire de l'Université de Sofia, 34 (1) (1937–38), 1–34.

**Ostrowski, A.**

1. *Ueber quasi-analytische Funktionen und Bestimmtheit asymptotischer Entwickelungen.* Acta Mathematica, 53 (1929), 181–266.

**Padé, H.**

1. *Sur la représentation approchée d'une fonctions par des fractions rationelles.* Annales Scientifiques de l'Ecole Normale Supérieure, (3) 9 (Supplement) (1892), S3–93.

2. *Sur le développement en fraction continue de la fonction F (h, l, h', u) et la généralisation de la théorie des fonctions sphériques.* Comptes Rendus, 141 (1905), 819–821.

3. *Recherches sur la convergence des développements en fractions continues d'une certaine catégorie des fonctions.* Annales Scientifiques de l'Ecole Normale Supérieure, (3), 24 (1907), 341–400.

**Paley, R. E. A. C. and Wiener, N.**

1. *Fourier Transforms in the Complex Domain.* American Mathematical Society Colloquium Publications, 19 (1934), 184 pp.

**Perron, O.**

1. *Die Lehre von den Kettenbrüchen.* Teubner, Leipzig, 1913, 520 pp.

2. *Erweiterung eines Markoffschen Satzes über die Konvergenz gewisser Kettenbruche.* Mathematische Annalen, 74 (1913), 545–554.

**Pincherle, S.**

1. *Sur les fractions continues algébriques.* Annales Scientifiques de l'Ecole Normale Supérieure, (3) 6 (1889), 145–152.

2. *Sulla generalizzazione delle frazioni continue algebriche.* Annali di Matematica, (2) 19 (1891), 75–95.

**Pólya, G.**

1. *Ueber den zentralen Grenzwertsatz der Wahrscheinlichkeitrechnung.* Mathematische Zeitschrift, 8 (1920), 171–181.

2. *Ueber gewisse notwendige Determinantkriterien für die Fortsetzbarkeit einer Potenzreihe.* Mathematische Annalen, 99 (1928), 687–706.

3. *Sur l'indétermination d'un théorème voisin du problème des moments.* Comptes Rendus, 207 (1938), 708–711.

**Pólya, G. und Szegö, G.**

1. *Aufgaben und Lehrsätze aus der Analysis*; v. I, II. Springer, Berlin, 1925.

**Possé, C.**

1. *Sur quelques applications des fractions continues algébriques.* St.-Petersburg, 1886, 175 pp.

**Riesz, F.**

1. *Sur certaines systèmes singulières d'équations intégrales.* Annales Scientifiques de l'Ecole Normale Supérieure, 28 (1911), 33–62.

2. *Sur l'approximation des fonctions continues et des fonctions sommables.* Bulletin of the Calcutta Mathematical Society, 20 (1930), 55–58.

**Riesz, M.**

1. *Sur le problème des moments. Première Note.* Arkiv för matematik, astronomi och fysik, 16 (12) (1921), 23 pp.

2. *Sur le problème des moments. Deuxième Note.* Arkiv för matematik, astronomi och fysik, 16 (19) (1922), 21 pp.

3. *Sur le problème des moments. Troisième Note.* Arkiv för matematik, astronomi och fysik, 17 (16) (1923), 52 pp.

4. *Sur le problème des moments et le théorème de Parseval correspondant.* Acta Litterarum ac Scientiarum (Szeged), 1 (1922–23), 209–225.

**Saks, S.**

1. *Theory of the integral.* Warszawa-Lwow, 1937, 344 pp.

**Schoenberg, I. J.**

1. *On a finite and infinite completely monotonic sequence.* Bulletin of the American Mathematical Society, 38 (1932), 72–76.

2. *On finite-rowed systems of linear inequalities in infinitely many variables.* Transactions of the American Mathematical Society, 34 (1933), 594–619.

**Schoenberg, I. J. and Hildebrandt, T. H.** (See Hildebrandt, T. H.)

**Schwetzow, K. I.**

1. *Ueber das Momentenproblem von Hamburger bei den zusätzlichen Bedingung das ein gegebenes Intervalle frei von Massen sei.* Communications de l'Institut des Sciences Mathématiques et Mécaniques de l'Université de Kharkoff et de la Société Mathematique de Kharkoff, (4), 16 (1940), 121–128.

**Sherman, J.**

1. *On the numerators of the convergents of the Stieltjes continued fraction.* Transactions of the American Mathematical Society, 35 (1933), 64–87.

**Sherman, J. and Shohat, J.**

1. *On the numerators of the continued fraction* $\frac{\lambda_1 |}{| x - c_1} - \frac{\lambda_2 |}{| x - c_2} - \ldots$ Proceedings of the National Academy of Sciences, 18 (1932), 283–287.

**Shohat, J. (Jacques Chokhate).**

1. *Sur le développement de l'intégrale* $\int_a^b \frac{p(y)\,dy}{x - y}$ *en fraction continue et sur les polynomes de Tchebycheff.* Rendiconti del Circolo Matematico di Palermo, 47 (1923), 25–46.

2. *The theory of closure of Tchebycheff polynomials for an infinite interval.* Bulletin of the American Mathematical Society, 30 (1924), 505–510.

3. *Sur les quadratures mécaniques et sur les zéros des polynomes de Tchebycheff dans un intervalle infini.* Comptes Rendus, 185 (1927), 597–598.

4. *On the continued fraction associated with and corresponding to the integral* $\int_a^b \frac{p(y)\,dx}{x - y}$. American Journal of Mathematics, 55 (1933), 218–230.

5. *Théorie générale des polynomes orthogonaux de Tchebichef.* Mémorial des Sciences Mathématiques, Fasc. 66 (1934), 69 pp.

6. *The relation of the classical orthogonal polynomials to the polynomials of Appell.* American Journal of Mathematics, 58 (1936), 453–464.

7. *On mechanical quadratures, in particular, with positive coefficients.* Transactions of the American Mathematical Society, 42 (1937), 461–496.

——— **and Fréchet, M.** (See Fréchet)

——— **and Sherman, J.** (See Sherman)

**Sleshinsky, N.**

1. *On convergence of continued fractions* (in Russian). Mémoires de la Société de Naturalistes de Nouvelle Russie, 10 (1889), 201–256.

**Smith, C.**

1. *On generalized Tchebycheff inequalities in Mathematical Statistics.* American Journal of Mathematics, 52 (1930), 109–126.

**Sonine, N.**

1. *On the approximate evaluation of definite integrals and on the related integral functions* (in Russian). Warshawskia Universitetskia Izwestia (Annals of the University of Varsaw), 1 (1887), 1–32.

2. *Sur l'intégrale* $\int_a^b F(x) \frac{dx}{z - x}$. Mémoires de l'Académie des Sciences de St.-Petersburg, classe physico-mathématique, (7) 38 (14) (1888), 44 pp.

3. *Sur quelques inégalités concernant les intégrales définies.* Mémoires de l'Académie des Sciences de St.-Petersburg, (8) 6 (6) 1898, 54 pp.

**Spencer, Vivian E.**

1. *Persymmetric and Jacobi determinant expressions for orthogonal polynomials.* Duke Mathematical Journal, 5 (1939), 333–356.

**Steinhaus, H. and Kaczmarz, S.** (See Kaczmarz, S.)

**Stekloff, W.**

1. *Quelques applications nouvelles de la théorie de fermeture au problème de représentation approchée des fonctions et au problème des moments.* Mémoires de l'Académie des Sciences de St.-Petersburg, (8) 32 (4) (1914), 74 pp.

2. *Application de la théorie de fermeture à la solution de certaines questions qui se rattachent au problème des moments.* Mémoires de l'Académie des Sciences de St.-Petersburg, (8) 33 (9) (1915), 52 pp.

3. *Sur le calcul approché des intégrales definies à l'aide des quadratures dites mécaniques.* Bulletin de l'Académie des Sciences de St.-Petersburg, (6) 10 (1916), 169–186.

4. *Sopra la teoria delle quadrature dette mechaniche.* Reale Accademia (nazionale) dei Lincei, Rendiconti della Classe di Scienze fisiche, matematiche et naturali, (5) 32 (1) (1923), 320–326.

5. *Sur les problèmes de représentation des fonctions à l'aide des polynomes, du calcul approché des intégrales définies, du développement des fonctions en séries infinies suivant les polynomes et de l'interpolation considérés au point de vue des idées de Tchebycheff.* Proceedings of the International Mathematical Congress, Toronto, 1 (1924), 631–640.

**Stieltjes, T. J.**

1. *Sur l'évaluation approchée des intégrales.* Comptes Rendus, 97 (1884), 740–742, 798–799.

2. *Sur un développement en fraction continue.* Comptes Rendus, 99 (1884), 508–509.

3. *Quelques recherches sur les quadratures dites mécaniques.* Annales Scientifiques de l'Ecole Normale Supérieure, (3) 1 (1884), 409–426.

4. *Recherches sur les fractions continues.* Mémoires presentés par divers Savants à l'Académie des Sciences de l'Institut de France, Sciences et mathématiques, (2) 32 (2) (1892), 196 pp.

5. *Recherches sur les fractions continues.* Annales de la Faculté des Sciences de Toulouse, (1) 8 (1894), T 1–122, (1) 9 (1895) A5–47.

6. *Correspondance d'Hermite et de Stieltjes.* 1 (48) (1884), 98–99; 1 (52) (1884), 107–114; 1 (167) (1889), 332–343; 1 (172) (1889), 348–352; 1 (175) (1889), 359–362; 2 (293) (1891), 142–143; 2 (295) (1891), 144–147; 2 (314) (1891), 185–187; 2 (316) (1891), 191–194; 2 (325) (1892), 216–222; 2 (328) (1892), 228–232; 2 (351) (1892), 270–276; 2 (383) 332–334; 2 (385) (1893), 337–339; 2 (386) (1893), 339–340; 2 (426) (1894), 434–435.

**Stone, M. H.**

1. *Linear transformations in Hilbert space and their applications to analysis.* American Mathematical Society Colloquium Publications, 15, New York, 1932, 622 pp.

**Stridsberg, E.**

1. *Några elementära undersökningar rörande fakultäter och deras arithmetika egenskaper. Not I.* Arkiv för Matematik, astronomi och fysik, 11 (25) (1917), 52 pp.

2. *Några aritmetiska undersökningar rörande fakulteter och vissa allmännare koefficientswiter. Not II.* Arkiv för matematik, astronomi och fysik, 13 (25) (1918), 70 pp.

3. *Några aritmetiska undersökningar rorande faculteter och vissa allmännare koefficientsviter. Not III.* Arkiv för matematik, astronomi och fysik, 15 (22) (1921), 126 pp.

**Szász, O.**

1. *Bemerkungen zu Herrn Perrons Erweiterunz eines Markoffschen Satzes über die Konvergenz gewisser Kettenbrüche.* Mathematische Annalen, 76 (1915), 301–314.

**Szegö, G.**

1. *A Hankel-fele formakrol.* Mathematikai ès termeszettudommanyi Ertesitò, 36 (1918), 497–538.

2. *Orthogonal Polynomials.* American Mathematical Society Colloquium Publications, 23, New York, 1939, 401 pp.

**Szegö, G. and Pólya, G.** (See Pólya, G.)
**Tamarkin, J. and Hille, E.** (See Hille, E.)
**Tamarkin, J. D.**
1. *On a theorem of S. Bernstein-Widder.* Transactions of the American Mathematical Society, 33 (1932), 893–896.
**Tartler, A.**
1. *On a certain class of orthogonal polynomials.* American Journal of Mathematics, 57 (1935), 627–644.
**Tchebycheff, P.**
1. *Sur les valeurs limites des intégrales.* Journal des Mathématiques pures et appliquées, (2) 19 (1874), 157–160; Collected Papers, 2, No. 10 (1907), 181–185.
2. *Sur la représentation des valeurs limites des intégrales par des residus intégraux.* Acta Mathematica, 9 (1887), 35–56.
3. *Sur les résidus intégraux qui donnent les valeurs approchées des intégrales.* Acta Mathematica, 12 (1889), 287–322.
**Thijssen, W. P.**
1. *Over het ontwikkelen eener functie van twee reëele veranderlijken in reeksen van veeltermen.* Verslagen van de Zittingen der wis—en natuurkundige afdecling dev Koninklijke Akademie van Weteschappen, 35 (2) (1926), 1100–1111.
2. *On the development of a function of two real variables into a series of polynomials.* Koninklijke (Nederlandsche) Akademie van Wetenschappen te Amsterdam. Proceedings of the Section of Science, 30 (1) (1927), 69–80.
**Titchmarsh, E. C.**
1. *Introduction to the theory of Fourier integrals.* Oxford, 1937, 390 pp.
2. *The theory of functions.* Oxford University Press, 1939, 2nd ed., 454 pp.
**Toeplitz, E. und Hellinger, E.** (See Hellinger, E.)
**Turán, P. and Erdös, P.** (See Erdös, P.)
**Uspensky, J.**
1. *Sur la convergence de quadratures, dites mécaniques, entre les limites infinies* (in Russian). Bulletin de l'Académie des Sciences de St.-Petersburg, (6) 10 (1916), 851–866.
2. *On the convergence of quadrature formulas related to an infinite interval.* Transactions of the American Mathematical Society, 30 (1928), 542–559.
**Van Vleck, E. B.**
1. *On an extension of the 1894 memoir of Stieltjes.* Transactions of the American Mathematical Society, 4 (1903), 297–332.
2. *Selected topics in the theory of divergent series and of continued fractions.* American Mathematical Society Colloquium Publications, 1, Boston Colloquium (1903), New York, 1905, 75–187.
**Verblunsky, S.**
1. *Some theorems on moment constants and functions.* Proceedings of the London Mathematical Society, (2) 37 (5) (1932), 338–382.
2. *On positive harmonic functions: a contribution to the algebra of Fourier series.* Proceedings of the London Mathematical Society, (2) 38 (1934), 125–157.
3. *Solution of a moment problem for bounded functions.* Proceedings of the Cambridge Philosophical Society, 32 (1936), 30–39.
4. *On the Fourier constants of bounded functions.* Proceedings of the Cambridge Philosophical Society, 32 (1936), 201–211.
5. *On the parametric representation of bounded functions.* Proceedings of the Cambridge Philosophical Society, 32 (1936), 521–529.
**Voronovskaia, E.**
1. *Ein Minimalproblem in der Theorie der Momentfolgen und die Abschätzung der Polynome.* Comptes Rendus de l'Académie des Sciences de l'URSS, (2) 4 (1) (1936), 103–106; (2) 4 (3) (1936), 147–150.
**Wald, A.**
1. *Limits of a distribution function determined by absolute moments and inequalities satisfied by absolute moments.* Transactions of the American Mathematical Society, 46 (1939), 280–306.

**Wall, H. S.**

1. *On the Padé approximants associated with the continued fraction and series of Stieltjes.* Transactions of the American Mathematical Society, 31 (1929), 91–116.

2. *On the extended Stieltjes series.* Transactions of the American Mathematical Society, 31 (1929), 771–781.

3. *On the Padé approximants associated with a positive definite power series.* Transactions of the American Mathematical Society, 33 (1931), 511–532.

4. *General theorems on the convergence of sequences of Padé approximants.* Transactions of the American Mathematical Society, 34 (1932), 409–416.

5. *Continued fractions and totally monotone sequences.* Transactions of the American Mathematical Society, 48 (1940), 105–184.

**Wall, H. S. and Garabedian, H. L.** (See Garabedian, H. L.)

**Wall, H. S., Garabedian, H. L. and Hille, E.** (See Garabedian, H. L.)

**Widder, D. V.**

1. *On the changes of sign of the derivatives of a function defined by a Laplace integral.* Proceedings of the National Academy of Sciences, 18 (1932), 112–114.

2. *Necessary and sufficient conditions for the representation of a function as a Laplace integral.* Transactions of the American Mathematical Society, 33 (1932), 851–892.

3. *The inversion of the Laplace integral and the related moment problem.* Transactions of the American Mathematical Society, 36 (1934), 107–200.

4. *An application of Laguerre polynomials.* Duke Mathematical Journal, 1 (1935), 126–136.

5. *The iterated Stieltjes transform.* Proceedings of the National Academy of Sciences, 23 (1937), 242–244.

6. *The Laplace Transform.* Princeton University Press, Princeton, 1941, 406 pp.

**Widder, D. V. and Wiener, N.**

1. *Remarks on the classical inversion formula for the Laplace integrals.* Bulletin of the American Mathematical Society, 44 (1938), 573–575.

**Wiener, N. and Paley, R. E. A. C.** (See Paley, R. E. A. C.)

**Wiener, N. and Widder, D. V.** (See Widder, D. V.)

**Wigert, S.**

1. *Sur les polynomes orthogonaux et l'approximation des fonctions continues.* Arkiv för matematik, astronomi och fysik, 17 (18) (1923), 15 pp.

**Wilson, R.**

1. *An extension of Hadamard-Pólya determinantal criteria for uniform functions.* Proceedings of the London Mathematical Society, (2) 39 (1935), 363–371.

**Wintner, A.**

1. *Ein qualitatives Kriterium für die Konvergenz des assoziierten Kettenbrüche.* Mathematische Zeitschrift, 30 (1929), 285–289.

2. *Spektraltheorie der unendlichen Matrizen*, Leipzig, 1929, 280 pp.

**Wintner, A. and Haviland, E. K.** (See Haviland, E. K.)

**Zygmund, A.**

1. *Trigonometrical Series.* Warszawa-Lwow, 1935, 331 pp.

# SUPPLEMENTARY BIBLIOGRAPHY

**Achyeser (Achieser), N. I.**

2. *The general theory of the polynomials of P. L. Chebyshev* (in Russian). Nauchnoe nasledie P. L. Chebysheva, vypusk pervyĭ: Matematika, Moscow-Leningrad, 1945, pp. 5–42.

**Agmon, S.**

1. *Sur deux théorèmes de M. S. Mandelbrojt*. Comptes Rendus Hebdomadaires des Séances de l'Académie des Sciences, Paris 228 (1949), 1835–1837.

**Bang, T.**

1. *A moment problem* (in Danish). Matematisk Tidsskrift B. (1946), 77–82.

**Belz, M. H.**

1. *Note on the Liapounoff inequality for absolute moments*. Annals of Mathematical Statistics 18 (1947), 604–605.

**Bernstein, S.**

3. *Constructive theory of functions as a development of Tchebychef's ideas* (in Russian with English summary). Bulletin de l'Académie des Sciences de l'URSS, Série Mathématique 9 (1945), 145–158.

**Boas, R. P.**

6. *Density theorems for power series and complete sets*. Transactions of the American Mathematical Society 61 (1947), 54–68.

7. *Sur les suites vérifiant des inégalités portant sur leurs différences*. Comptes Rendus Hebdomadaires des Séances de l'Académie des Sciences, Paris 224 (1947), 1683–1685.

**Bochner, S. and Fan, K.**

1. *Distributive order-preserving operations in partially ordered vector sets*. Annals of Mathematics (2) 48 (1947), 168–179.

**Churchill, E.**

1. *Information given by odd moments*. Annals of Mathematical Statistics 17 (1946), 244–246.

**Eidelheit, M.**

1. *Quelques remarques sur les fonctionelles linéaires*. Studia Mathematica 10 (1948), 140–147.

**Fan, K.**

1. *On positive definite sequences*. Annals of Mathematics (2) 47 (1946), 593–607.

**Fan, K. and Bochner, S.** (See Bochner, S.)

**Fortet, R.**

1. *Calcul des moments d'une fonction de répartition à partir de sa caractéristique*. Bulletin des Sciences Mathématiques (2) 68 (1944), 117–131.

**Fuchs, W. H. J.**

1. *On a generalization of the Stieltjes moment problem*. Bulletin of the American Mathematical Society 52 (1946), 1057–1059.

**Geronimus, J.**

7. *On polynomials orthogonal on the circle, on trigonometric moment problem and on allied Carathéodory and Schur functions*. Comptes Rendus (Doklady) de l'Académie des Sciences de l'URSS N.S. 39 (1943), 291–295.

8. *On Gauss' and Tchebycheff's quadrature formulas*. Bulletin of the American Mathematical Society 50 (1944), 217–221.

9. *On polynomials orthogonal on the circle, on trigonometric moment-problem and on allied Carathéodory and Schur functions*. Matematicheskii Sbornik N.S. 15 (57) (1944), 99–130.

10. *On some distribution functions connected with systems of polynomials*. Comptes Rendus (Doklady) de l'Académie des Sciences de l'URSS N.S. 44 (1944), 355–359.

11. *On the trigonometric moment problem*. Annals of Mathematics (2) 47 (1946), 742–761.

**Ghizzetti, A.**

1. *Ricerche sui momenti di una funzione limitata compresa fra limiti assegnati*. Atti della Reale Accademia d'Italia, Memorie della Classe di Scienze Fisiche, Matematiche e Naturali (7) 13 (1942), 1165–1199.

2. *Sui momenti di una funzione limitata*. Atti della Reale Accademia della Scienze di Torino, Classe di Scienze Fisiche, Matematiche e Naturali 77 (1942), 198–208.

3. *Sui momenti di 2° ordine di una legge di probabilità in n dimensioni*. Regia Università di Roma e Reale Istituto Nazionale di Alta Matematica, Rendiconti di Matematica e delle sue Applicazione (5) 4 (1943), 94–101.

4. *Condizioni necessarie e sufficienti per i momenti di una funzione limitata*. Atti dell'Accademia Nazionale dei Lincei, Rendiconti della Classe di Scienze Fisiche, Matematiche e Naturali (8) 2 (1947), 533–536.

5. *Sul problema dei momenti*. Università e Politecnico di Torino. Rendiconti del Seminario Matematico e Fisico 8 (1947–49), 93–108.

**Greenstein, D. S.**

1. *On a criterion for determinate moment sequences*, Proc. Amer. Math. Soc. 13 (1962), 120-121.

**Hagstroem, K. G.**

1. *Un problème de calcul stochastique*. Försäkringsmatematiska studier tillägnade Filip Lundberg, Stockholm, 1946, pp. 104–127.

**Hamburger, H.**

4. *Hermitian transformations of deficiency-index (1,1), Jacobi matrices and undetermined moment problems*. American Journal of Mathematics 66 (1944), 489–522.

**Hellinger, E. D. and Wall, H. S.**

1. *Contributions to the analytic theory of continued fractions and infinite matrices*. Annals of Mathematics (2) 44 (1943), 103–127.

**Horváth, J.**

1. *Sur l'approximation polynomiale des fonctions sur une demi-droite*. Comptes Rendus Hebdomadaires des Séances de l'Académie des Sciences, Paris 227 (1948), 1074–1076.

**Julia, G.**

1. *Remarques géométriques sur le problème des moments dans l'espace hilbertien*. Comptes Rendus Hebdomadaires des Séances de l'Académie des Sciences, Paris 216 (1943), 257–260.

**Karlin, S. and Shapley, L. S.**

1. *Geometry of reduced moment spaces*. Proceedings of the National Academy of Sciences of the United States of America 35 (1949), 673–677.

**Krein, M.**

4. *On a remarkable class of Hermitian operators*. Comptes Rendus (Doklady) de l'Académie des Sciences de l'URSS N.S. 44 (1944), 175–179.

5. *On a generalized problem of moments*. Comptes Rendus (Doklady) de l'Académie de Sciences de l'URSS N.S. 44 (1944), 219–222.

6. *Infinite J-matrices and matrices of the moment problem* (in Russian). Doklady Akademii Nauk SSSR N.S. 69 (1949), 125–128.

**Krein, M. and Krasnoselskii, M. A.**

1. *Fundamental theorems on the extension of Hermitian operators and certain of their applications to the theory of orthogonal polynomials and the problem of moments* (in Russian). Uspehi Matematicheskih Nauk N.S. 2, no. 3 (19) (1947), 60–106.

**Krein, M. and Šmulian, V.**

1. *On regularly convex sets in the space conjugate to a Banach space*. Annals of Mathematics (2) 41 (1940), 556–583.

**Lifshetz, M. (Livshitz, M.)**

2. *On an application of the theory of Hermitian operators to the generalized problem of moments*. Comptes Rendus (Doklady) de l'Académie des Sciences de l'URSS N.S. 44 (1944), 3–7.

**Loève, M.**

1. *Remarques sur la majoration de certaines transformées*. Comptes Rendus Hebdomadaires des Séances de l'Académie des Sciences, Paris 225 (1947), 31–33.

**Mandelbrojt, S.**

1. *Quelques considérations sur le problème des moments*. Comptes Rendus Hebdomadaires des Séances de l'Académie des Sciences, Paris 226 (1948), 862–864.

2. *Théorèmes d'unicité*. Annales Scientifiques de l'École Normale Supérieure (3) 65 (1948), 101–138.

**Neumark, M. A.**

1. *On extremal spectral functions of a symmetric operator*. Comptes Rendus (Doklady) de l'Académie des Sciences de l'URSS N.S. 54 (1946), 7–9.

**Rees, M. R.**

1. *A theorem of three moments*. Journal of the London Mathematical Society 22 (1947), 124–130.

**Scott, W. T.**

1. *The corresponding continued fraction of a J-fraction*. Annals of Mathematics (2) 51 (1950), 56–67.

**Selberg, H. L.**

1. *On an inequality in mathematical statistics* (in Norwegian). Norsk Matematisk Tidsskrift 24 (1942), 1–12.

**Šmulian, V. and Krein, M.** (See Krein, M.)

**Thron, W. J.**

1. *Convergence regions for the general continued fraction*. Bulletin of the American Mathematical Society 49 (1943), 913–916.

**Verblunsky, S.**

6. *Two moment problems for bounded functions*. Proceedings of the Cambridge Philosophical Society 42 (1946), 189–196.

7. *On the initial moments of a bounded function*. Proceedings of the Cambridge Philosophical Society 43 (1947), 275–279.

8. *On the moments of a concave function*. Journal of the London Mathematical Society 22 (1947), 115–120.

9. *Additional note on two moment problems for bounded functions*. Proceedings of the Cambridge Philosophical Society 44 (1948), 140–142.

10. *On a problem of moments*. Proceedings of the Cambridge Philosophical Society 45 (1949), 1–4.

**Wall, H. S.**

6. *Analytic theory of continued fractions*. Van Nostrand, New York, 1948.

**Wall, H. S. and Hellinger, E. D.** (See Hellinger, E. D.)

**Wintner, A.**

3. *The moment problem of enumerating distributions*. Duke Mathematical Journal 12 (1945), 23–25.

**Zygmund, A.**

2. *A remark on characteristic functions*. Annals of Mathematical Statistics 18 (1947), 272–276.

# INDEX

(Numbers refer to pages)

FGHIJ–AMS–89876543